VOLUME FIFTY THREE

ADVANCES IN CHEMICAL ENGINEERING

Bridging Scales in Modelling and Simulation of Non-Reacting and Reacting Flows. Part II

ADVANCES IN CHEMICAL ENGINEERING

VOLUME FIFTY THREE

ADVANCES IN CHEMICAL ENGINEERING

Bridging Scales in Modelling and Simulation of Non-Reacting and Reacting Flows. Part II

Edited by

ALESSANDRO PARENTE

Combustion and Robust Optimisation (BURN) research group
Aero-Thermo-Mechanics Department
Université Libre de Bruxelles
Bruxelles, Belgium

JURAY DE WILDE

Institute of Mechanics, Materials and Civil Engineering (iMMC)
Materials & Process Engineering (IMAP)
Université catholique de Louvain
Louvain-la-Neuve, Belgium

ELSEVIER

ACADEMIC PRESS

An imprint of Elsevier

Academic Press is an imprint of Elsevier
50 Hampshire Street, 5th Floor, Cambridge, MA 02139, United States
525 B Street, Suite 1800, San Diego, CA 92101-4495, United States
The Boulevard, Langford Lane, Kidlington, Oxford OX5 1GB, United Kingdom
125 London Wall, London, EC2Y 5AS, United Kingdom

First edition 2018

Notices
Knowledge and best practice in this field are constantly changing. As new research and experience broaden our understanding, changes in research methods, professional practices, or medical treatment may become necessary.

Practitioners and researchers must always rely on their own experience and knowledge in evaluating and using any information, methods, compounds, or experiments described herein. In using such information or methods they should be mindful of their own safety and the safety of others, including parties for whom they have a professional responsibility.

To the fullest extent of the law, neither the Publisher nor the authors, contributors, or editors, assume any liability for any injury and/or damage to persons or property as a matter of products liability, negligence or otherwise, or from any use or operation of any methods, products, instructions, or ideas contained in the material herein.

ISBN: 978-0-12-815094-8
ISSN: 0065-2377

Publisher: Zoe Kruze
Acquisition Editor: Sam Mahfoudh
Editorial Project Manager: Shellie Bryant
Production Project Manager: Vignesh Tamil
Cover Designer: Greg Harris

Typeset by SPi Global, India

CONTENTS

CONTRIBUTORS

Maike W. Baltussen
Multiphase Reactors Group, Eindhoven University of Technology, Eindhoven, The Netherlands

Sofiane Benyahia
National Energy Technology Laboratory, Morgantown, WV, United States

Kay A. Buist
Multiphase Reactors Group, Eindhoven University of Technology, Eindhoven, The Netherlands

Johannes A.M. Kuipers
Multiphase Reactors Group, Eindhoven University of Technology, Eindhoven, The Netherlands

Liqiang Lu
National Energy Technology Laboratory, Morgantown, WV, United States

Federico Municchi
Institute of Process and Particle Engineering, Graz University of Technology, Graz, Austria

Elias A.J.F. Peters
Multiphase Reactors Group, Eindhoven University of Technology, Eindhoven, The Netherlands

Stefan Radl
Institute of Process and Particle Engineering, Graz University of Technology, Graz, Austria

PREFACE

In the context of the International Francqui Chair of Rodney O. Fox, Professor at Iowa State University, two volumes of *Advances in Chemical Engineering* are dedicated to "Bridging Scales in Modelling and Simulation of Reacting Flows." As Professor Rodney Fox, the contributors to this issue have made numerous ground-breaking contributions to the field of multiphase and reactive flow modelling. The impact of the advances made extends far beyond chemical engineering and touches every technological area dealing with turbulent flow and chemical reactions (e.g., combustion, atmospheric science, nuclear fuel processing).

All contributions to this collection recognize that future scientific and technological developments will greatly benefit from detailed simulations and that a fundamental approach combining numerical simulations, experiments, and rigorous validation needs to be developed. Scale-bridging approaches are essential for simulations at industrial scale as they allow using the information gathered at the microscale for more accurate calculations at the macroscale. This is not a trivial task, and implies a thorough understanding of the physical processes involved and their interactions, as well as the possible degree of simplification of the problem, in view of the multidisciplinary application.

Whereas Volume 52 addresses the coupling between flow and species composition fields, Volume 53 focuses on fluidized bed reactors and the advances made in developing scale-bridging approaches. The methods presented in the different chapters are illustrated with industrially relevant applications.

Alessandro Parente and Juray De Wilde

CHAPTER ONE

Multiscale Modelling of Dense Gas–Particle Flows

Maike W. Baltussen, Kay A. Buist, Elias A.J.F. Peters, Johannes A.M. Kuipers[1]
Multiphase Reactors Group, Eindhoven University of Technology, Eindhoven, The Netherlands
[1]Corresponding author: e-mail address: J.A.M.Kuipers@tue.nl

Contents

Abstract

In large-scale industrial processes involving granulation, coating, and production of base chemicals and polymers dense particulate flows with coupled mass, momentum, and heat transfer are frequently encountered. Both (effective) fluid–particle and (dissipative) particle–particle interactions need to be accounted for because the mutual competition between these phenomena govern the key features of dense gas–particle flows such as regime transitions. These interactions prevail at different length scales and consequently a multiscale approach is adopted to arrive at a quantitative description of these complex flows. In this approach detailed models, taking into account the relevant details of fluid–particle interaction (DNS) and particle–particle interaction (DEM) are used to obtain closure laws to feed two-fluid models (TFMs) which can be used to simulate the flow on a much larger (industrial) scale. In this chapter, we will discuss recent advances in the multiscale simulation of dense gas-fluidized beds. The governing

Advances in Chemical Engineering, Volume 53
ISSN 0065-2377
https://doi.org/10.1016/bs.ache.2018.02.001

equations will be presented as well as the key features of the numerical solution method. For each model type, illustrative computational results will be presented. Finally, areas which need substantial further attention will be discussed.

ABBREVIATIONS

CFD computational fluid dynamics
CFM continuous forcing method
CPU central processing unit
CSTR continuously stirred tank reactor
DFM direct forcing method
DIA digital image analysis
DNS direct numerical simulation
DPM discrete particle model
DEM discrete element model
EL Euler–Lagrange
FOU first-order upwind
ICGM incomplete conjugate gradient method
IBM immersed boundary method
KTGF kinetic theory of granular flow
LBM Lattice Boltzmann method
LLDPE linear low-density polyethylene
MD molecular dynamics
ODE ordinary differential equation
PDF probability density function
PEPT positron emission particle tracking
PIV particle image velocimetry
PRS particle resolved simulation
STA simultaneous thermal analysis
TFM two-fluid model
TGA thermogravimetric analysis
TVD total variation diminishing
UPW upwind

NOTATION

$C_{i,f}$ concentration of species i (mole/m^3)
$\langle C_{A,f} \rangle$ flow-average concentration of species A (mole/m^3)
C_d drag coefficient (−)
C_p heat capacity (J/(kg K))
C_m net convective mass flux (mole/(m^3 s))
C_h net convective heat flux (W/m^3)
d_p particle diameter (m)
D mass defect according to the continuity equation (kg/m^3)
$D(\bar{r})$ distribution function (−)
$D_{i,f}$ diffusivity of species i in the fluid (m^2/s)
e_n normal restitution coefficient (−)

F dimensionless drag coefficient (−)
g_0 radial distribution function (−)
I_p particle moment of inertia (kg m^2)
J_g Jacobian for pressure correction (s^2/m^2)
J_s Jacobian for voidage correction (kg/m^3)
k_m fluid–particle mass transfer coefficient (m/s)
m_p particle mass (kg)
Nu particle Nusselt number (−)
p, P pressure (N/m^2)
r radial coordinate (m)
Pr Prandtl number (−)
Re_p particle Reynolds number (−)
T temperature (K)
$\langle T_f \rangle$ flow-average temperature (K)
t time (s)
U superficial gas velocity (m/s)
z axial coordinate (m)

GREEK LETTERS

α specific interphase heat transfer coefficient (W/m^3)
α_p fluid–particle heat transfer coefficient (W/m^2)
β interphase momentum transfer coefficient (kg/(m^3 s))
Δt time step (s)
Δp_g pressure correction (N/m^2)
$\Delta \varepsilon_g$ voidage correction (−)
ΔH_r heat of reaction (J/mol)
ε volume fraction (−)
ϕ angle (−)
Φ specularity coefficient (−)
$\Phi_{m,f\rightarrow s}$ fluid–particle mass transfer rate (mole/s)
$\Phi_{h,f\rightarrow s}$ fluid–particle heat transfer rate (W)
$\overline{\Phi}_f$ fluid phase heat flux (W/m^2)
γ dissipation rate of fluctuation kinetic energy of the solid phase (kg/(m s^3))
κ thermal conductivity (W/(m K))
λ bulk viscosity (kg/(m s))
μ dynamic viscosity (kg/(m s))
$\overline{\omega}_p$ particle rotational velocity (s^{-1})
ρ density (kg/m^3)
θ granular temperature (m^2/s^2)

VECTORS

$\overline{A}$ momentum density (kg/(m^2 s))
$\overline{C}_m$ net convective momentum flux (N/m^3)

$\overline{f}_{f \to p}$ volumetric force exerted by the fluid on the particles (N/m^3)
$\overline{F}_{f \to s}$ fluid–particle force (N)
$\overline{g}$ gravitational acceleration (m/s^2)
$h_{f \to p}$ volumetric fluid–particle heat exchange rate (W/m^3)
$\overline{n}$ unit outward normal (–)
$\overline{q}_s$ pseudo Fourier heat flux (kg/s^3)
$\overline{u}$ velocity (m/s)
$\overline{w}_p$ particle translational velocity (m/s)
$\overline{T}_{f \to s}$ fluid–particle torque (N m)

TENSORS

$\boldsymbol{D}$ rate of deformation tensor (s^{-1})
$\boldsymbol{I}$ identity tensor (–)
$\boldsymbol{\tau}$ viscous stress tensor (N/m^2)

SUBSCRIPTS AND SUPERSCRIPTS

con convection
e effective
f static friction, fluid phase
g gas phase
mf minimum fluidization
p particle phase
s solid phase
vis viscous
w wall

OPERATORS

$\frac{\partial}{\partial t}$ partial time derivative (s^{-1})
∇ gradient operator (m^{-1})
$\nabla \cdot$ divergence operator (m^{-1})

1. INTRODUCTION

Dense gas–particle flows are very often encountered in the process industries involving physical (coating, drying, and granulation) and chemical (synthesis of base chemicals, gasification, and combustion) transformations.

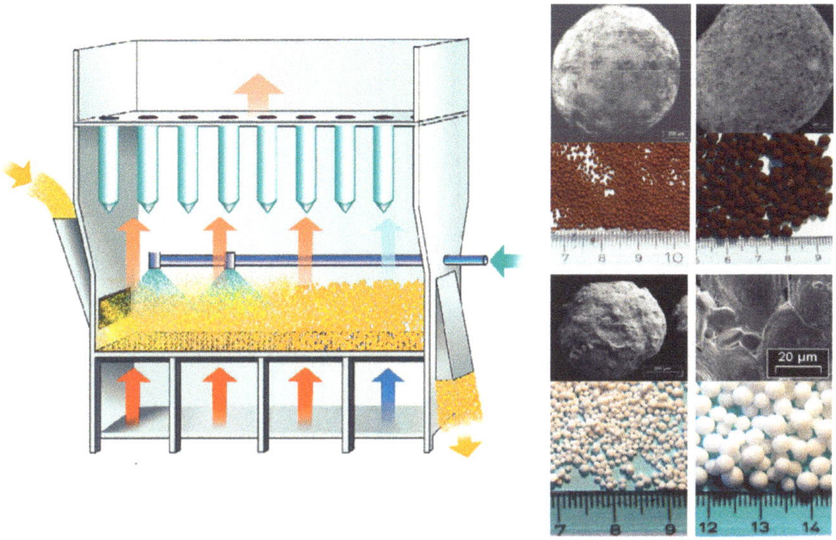

Fig. 1 Schematic representation of a gas-fluidized bed granulator (*left*) for the production of functional particles (solids inlet and outlet (*yellow*), hot (*red*) and cold (*dark blue*) gas inlet, and liquid binder inlet (*light blue*)). Examples of functional particles produced in this type of equipment are also shown (*right*).

In Fig. 1, shown as an example to highlight the complexity of such systems, a schematic representation of a gas-fluidized bed granulator for the production of functional particles is given. This apparatus consists of various sections with solids inlet and outlet, hot and cold gas inlets, and a liquid binder inlet facilitating top spraying of the bed. Binder deposition on the (primary) bed particles causes the growth of the particles to the desired size. Clearly in this unit operation a complex multiscale problem is encountered with poly-disperse dense gas–particle flow accompanied by coupled mass and heat transport. Moreover the introduction of the liquid leads to (partially) wet particles with complicated strongly dissipative particle–particle and particle–wall collisions (Buck et al., 2017; Tang et al., 2017). Such strongly dissipative collisions favor the formation of heterogeneous flow structures with considerable impact on the rate of mass, momentum, and heat exchange and hence on the performance of fluidized beds as multiphase chemical reactors. To further elucidate this aspect, we shown in Fig. 2 simulation results obtained from a hard sphere discrete particle model (DPM) for two-dimensional (2D) riser flow (see Hoomans, 2000).

From Fig. 2, it can be seen that the flow structure for the case with ideal particles produces a much more homogeneous particle distribution in

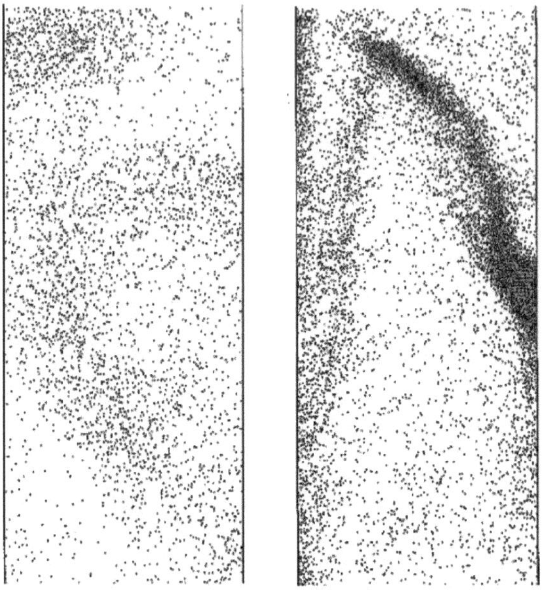

Fig. 2 Simulation results obtained from a hard sphere discrete particle model (DPM) for two-dimensional (2D) riser flow (Hoomans, 2000). Particle size and density: 500 μm and 2660 kg/m^3, superficial gas velocity: 5.0 m/s, solids flux 25 kg/(m^2 s), riser diameter: 0.08 m. *Left*: ideal collisions. *Right*: nonideal collisions. Here e and μ represent, respectively, the normal restitution coefficient and the tangential friction coefficient. Values for the collision parameters for particle–particle and particle–wall collisions are chosen the same.

comparison with the case where the particle collisions are nonideal. For the latter case, despite the fact that the particles are introduced uniformly at the bottom of the riser, due to the dissipative particle–particle and particle–wall collisions heterogeneous flow structures (i.e., so-called strands or clusters) emerge. The riser hydrodynamics is significantly influenced by the presence of strands or clusters and consequently the quality of the gas–solid contact is profoundly influenced and has been subject of extensive experimental and computational studies.

From these two examples, the enormous complexity of dense gas–particle flows is apparent and the associated huge challenge to develop predictive computational tools which can facilitate the optimization and design of gas-fluidized beds. In a broader sense, these tools could aid in improving the performance of plants processing solids that are reported to have been performing below their design expectations during the last decades

(Merrow, 1985). Owing to the continuous increasing computer capacity and rapid developments in computational science detailed simulation of complex multiphase flow phenomena has received a lot of attention during the last few decades.

The hydrodynamic approach to fluidization was started by Davidson (1961) through the analysis of single bubble motion in an infinite bed using two continuity equations and Darcy's law for porous media flow. He assumed irrotational solids flow around the rising gas bubble and his relatively simple model was able to capture the main features of the motion of isolated bubbles in dense gas-fluidized beds and laid the foundation for reactor models of dense gas-fluidized beds involving catalytic conversion (Kunii and Levenspiel, 1991).

Hydrodynamic modelling of dense gas-fluidized beds was pioneered by Gidaspow and coworkers using the full set of continuum equations (see review paper of Gidaspow, 1986) and demonstrated the abilities and potential of this approach as a new research and development tool for in depth study of gas-fluidized beds. Their pioneering work triggered the meanwhile massive research on hydrodynamic modelling of gas-fluidized beds and laid the foundation for the development of more advanced hydrodynamic models based on the kinetic theory of granular flow (KTGF) (see Ding and Gidaspow, 1990).

Their model was derived on basis of the Boltzmann equation for the velocity distribution of particles and can be classified as a generalization of the volume-averaged Navier–Stokes equations proposed by Anderson and Jackson (1967). Contrary to the classical TFM, the KTGF-based model provides closures for the solid phase transport coefficients (such as the shear and bulk viscosity) in terms of the granular temperature which is obtained by solving simultaneously a transport equation for the kinetic energy associated with the fluctuating solids motion. In the classical TFM, the values for the apparent viscosity are obtained from rheological measurements or indirect measurements based on bubble shape (Kuipers et al., 1992b).

Modern continuum models for dense gas-fluidized beds embedded in commercial CFD packages are typically based on the KTGF. Owing to the very complex phase interactions and the interplay with physical and/or chemical transformations, as evident from the two examples presented in this section, considerable challenges remain to arrive at a thorough fundamental understanding of these flows and development of design tools with significantly improved predictive capabilities. Owing to the fact that a very large variation in time and length scales exists in dense gas–particle flows a multiscale modelling approach has been adopted comprising several CFD models with different degree of sophistication.

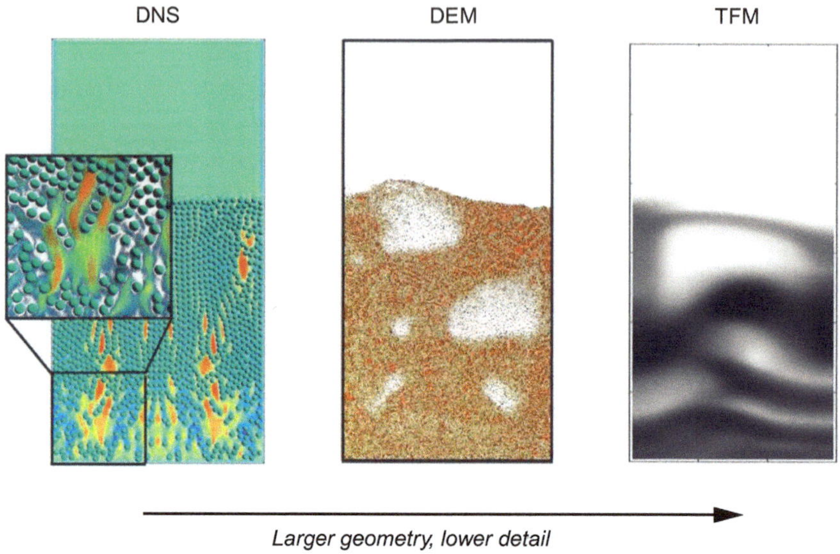

Fig. 3 Multiscale approach for dense gas–particle flows comprising direct numerical simulation (DNS), the discrete element model (DEM), and the two-fluid model (TFM) based on the kinetic theory of granular flow (KTGF). *Reproduced with permission from Deen NG, Kuipers JAM: Direct numerical simulation of fluid flow and mass transfer in dense fluid–particle systems,* Ind Eng Chem Res *52:11266–11274, 2013.*

This multiscale approach for dense gas–particle flows is described in detail by Van der Hoef et al. (2004, 2008) and comprises (see Fig. 3) direct numerical simulation (DNS), the discrete element model (DEM), and the two-fluid model (TFM) or continuum approach based on the KTGF. Subsequently the three levels of modelling comprising the multiscale approach presented in Fig. 3 will be discussed and presented in more detail.

2. DNS

In DNS or particle resolved simulation (PRS), a sufficiently refined computational grid is used to solve the governing equations and captures all relevant details of the flow field at a scale (much) smaller than the size of individual particles. The effective fluid–particle interaction (drag coefficient, heat and mass transfer coefficients) in dense particle swarms is evaluated as a function of the relevant parameters (solids volume fraction, particle Reynolds number, Prandtl (*Pr*) number, and Schmidt (*Sc*) number) and subsequently suitably parameterized. The Newtonian equations of motion are solved for the suspended particles while accounting for particle–particle and particle–wall collisions similar to the approach adopted in DEM. Additional equations for the particles such as species and thermal energy equations can be added with relative ease.

The underlying CFD methods are typically based on either fixed grids or body-conforming grids. Fixed grid methods such as the immersed boundary method (IBM) are preferred in case many moving particles need to be considered because the dynamic remeshing required in methods using body-conforming grids becomes very CPU demanding. Of course as an alternative the Lattice Boltzmann method (LBM) can be used as well. The IBM has gained considerable interest in recent years to handle flows in complex geometries due to the fact that this method is very CPU (and memory) efficient. Roughly two main types of IBM methods can be distinguished: continuous forcing methods (CFM) and direct forcing methods (DFM). In CFM, suitable source terms are added to the transport equations to enforce the boundary condition (for instance the Dirichlet condition) at the boundary of the immersed object, whereas in DFM this boundary condition is enforced at the level of the discretized transport equations. For details the interested reader is referred to excellent reviews by Peskin (2002), Mittal and Iaccarino (2005), and furthermore to Deen et al. (2012) and Deen and Kuipers (2013, 2014) detailing our IBM based on a ghost fluid approach.

2.1 Governing Equations DNS

The system of interest requires the simultaneous solution of the fluid phase and solid (particle) phase equations. For the model presented in this chapter, we make use of the following main assumptions:

- The fluid phase is incompressible and obeys the Newtonian stress closure model.
- All physical properties of the fluid phase and solids phase are constant.
- The solid phase consists of spherical catalyst particles where at the surface of the particles an infinitely fast chemical reaction involving a single chemical species A with associated heat effect takes place.
- Intraparticle temperature gradients are negligible.

2.1.1 Fluid Phase Equations

The transport phenomena in the fluid phase are governed by the conservation equations for mass, momentum, thermal energy, and chemical species, respectively, given by:

$$(\nabla \cdot \overline{u}) = 0 \tag{1}$$

$$\frac{\partial \rho_{\mathrm{f}} \overline{u}}{\partial t} + (\nabla \cdot \rho_{\mathrm{f}} \overline{u u}) = -\nabla p + \mu_{\mathrm{f}} \nabla^2 \overline{u} + \rho_{\mathrm{f}} \overline{g} \tag{2}$$

$$\rho_{\mathrm{f}} C_{\mathrm{p,f}} \left[\frac{\partial}{\partial t}(T_{\mathrm{f}}) + (\nabla \cdot \overline{u} T_{\mathrm{f}}) \right] = \kappa_{\mathrm{f}} \nabla^2 T_{\mathrm{f}} \tag{3}$$

$$\frac{\partial c_{i,\mathrm{f}}}{\partial t} + (\nabla \cdot c_{i,\mathrm{f}} \overline{u}) = D_{i,\mathrm{f}} \nabla^2 c_{i,\mathrm{f}} \quad i = 1, \ldots, N \tag{4}$$

In these equations, ρ_{f}, μ_{f}, $C_{\mathrm{p,f}}$, κ_{f} represent, respectively, the density, viscosity, heat capacity, and thermal conductivity of the fluid, whereas $D_{i,\mathrm{f}}$ represents the diffusivity of species "i" in the fluid. These equations are coupled to the particle equations which are given subsequently.

2.1.2 Particle Equations

The translational and rotational motion of the suspended solid particles is governed by the Newtonian equations of motion, respectively, given by:

$$m_{\mathrm{p}} \frac{\mathrm{d}\overline{w}_{\mathrm{p}}}{\mathrm{d}t} = m_{\mathrm{p}} \overline{g} + \overline{F}_{\mathrm{f} \to \mathrm{s}} \tag{5}$$

and

$$I_{\mathrm{p}} \frac{\mathrm{d}\overline{\omega}_{\mathrm{p}}}{\mathrm{d}t} = \overline{T}_{\mathrm{f} \to \mathrm{s}} \tag{6}$$

where m_{p} and I_{p} represent, respectively, the mass and the moment of inertia of the particle. The final terms on the right-hand sides in Eqs. (5) and (6) account for the fluid–particle interaction (respectively, drag and torque). The particle temperature is governed by (assuming uniform particle temperature) the following equation:

$$m_{\mathrm{p}} C_{\mathrm{p,p}} \frac{\mathrm{d}T_{\mathrm{p}}}{\mathrm{d}t} = \Phi_{\mathrm{h,\,f} \to \mathrm{s}} + \Phi_{\mathrm{m,\,f} \to \mathrm{s}}(-\Delta H_{\mathrm{r}}) \tag{7}$$

The first term on the right-hand side of Eq. (7) accounts for the heat transfer rate from the fluid to the particle, whereas the second term represents the heat source associated with a mass transfer limited exothermal surface reaction. The drag, torque, and the fluid–particle mass and heat transfer rates are computed once the solution of the fluid phase equations has been completed.

2.2 Numerical Solution Method DNS

The first step involves the time discretization of the momentum equation leading to the following expression:

$$\alpha\left[\rho_f \overline{u}^{n+1}\right] = \beta\left[\rho_f \overline{u}^{n}\right] + \gamma\left[\rho_f \overline{u}^{n-1}\right] - \Delta t \nabla p^{n+1} - \Delta t \overline{C}_m^n + \Delta t\left[\mu_f \nabla^2 \overline{u}^{n+1}\right] + \rho_f \overline{g} \Delta t \qquad (8)$$

where n indicates the time index and $\overline{C}_m$ the net convective momentum flux given by:

$$\overline{C}_m = \rho_f(\nabla \bullet \overline{uu}) \qquad (9)$$

For the discrete representation of the time derivative, we use a second-order three-point backward scheme (with $\alpha = 1.5, \beta = 2.0$, and $\gamma = -0.50$) requiring three levels of storage for the velocity components. For the first time step however level $(n - 1)$ is not available and consequently in this case: $\alpha = 1.0, \beta = 1.0$, and $\gamma = 0.0$ is used. For the solution of Eq. (8), a two-step projection method is used whereas a first step a tentative velocity field $\overline{u}^{**}$ is computed from:

$$\alpha\left[\rho_f \overline{u}^{**}\right] = \beta\left[\rho_f \overline{u}^{n}\right] + \gamma\left[\rho_f \overline{u}^{n-1}\right] - \Delta t \overline{C}_m^n + \Delta t\left[\mu_f \nabla^2 \overline{u}^{**}\right] + \rho_f \overline{g} \Delta t \qquad (10)$$

For the Laplace operator standard central second-order finite difference representations are used whereas the convection terms are evaluated with a second-order flux delimited Barton scheme (Centrella and Wilson, 1984). The final $(n+1)$ velocities are obtained from the following equation where the tentative (intermediate) velocity field is corrected to include the effect of the pressure gradient:

$$\overline{u}^{n+1} = \overline{u}^{**} - \frac{\Delta t}{\alpha \rho_f} \nabla p^{n+1} \qquad (11)$$

The pressure at the new time level is computed from the Poisson equation which can be obtained by taking the divergence of Eq. (11) and enforcing continuity on the final $(n+1)$ velocity field:

$$\frac{1}{\rho_f} \nabla^2 p^{n+1} = \frac{\alpha}{\Delta t}(\nabla \cdot \overline{u}^{**}) \qquad (12)$$

Once the solution of the momentum equations has been achieved the thermal energy equation and the species conservation equation are solved using the same temporal discretization scheme as applied for the momentum equation:

$$\alpha\left[\rho_f C_{p,f} T_f^{n+1}\right] = \beta\left[\rho_f C_{p,f} T_f^{n}\right] + \gamma\left[\rho_f C_{p,f} T_f^{n-1}\right] - \Delta t C_h^n + \Delta t\left[\kappa_f \nabla^2 T_f^{n+1}\right] \quad (13)$$

$$\alpha c_{A,f}^{n+1} = \beta c_{A,f}^{n} + \gamma c_{A,f}^{n-1} - \Delta t C_m^n + \Delta t\left[D_{A,f} \nabla^2 c_{A,f}^{n+1}\right] \quad (14)$$

where the net convective heat flux C_h^n and the net convective species molar flux C_m^n are given, respectively, by:

$$C_h^n = \rho_f C_{p,f} (\nabla \bullet \overline{u} T_f)^n \quad (15)$$

$$C_m^n = (\nabla \bullet \overline{u} c_{A,f})^n \quad (16)$$

For the spatial discretization of the convection and diffusion terms appearing in the thermal energy and species equations, we have used exactly the same schemes as for the solution of the momentum equations. For the solution of the algebraic equations resulting from the spatial discretization of Eqs. (10), (12), (13), and (14), we use a sparse matrix solver based on the incomplete conjugate gradient method (ICGM).

For the time integration of the solid phase equations (which we solve after the fluid phase equations), we use the trapezoidal rule which offers second-order accuracy and lead to the following equations:

$$\overline{w}_p^{n+1} = \overline{w}_p^{n} + \overline{g}\Delta t + \frac{\Delta t}{2m_p}\left[\overline{F}_{f\to s}^{n} + \overline{F}_{f\to s}^{n+1}\right] \quad (17)$$

$$\overline{\omega}_p^{n+1} = \overline{\omega}_p^{n} + \frac{\Delta t}{2I_p}\left[\overline{T}_{f\to s}^{n} + \overline{T}_{f\to s}^{n+1}\right] \quad (18)$$

$$T_p^{n+1} = T_p^n + \frac{\Delta t}{2m_p C_{p,p}}\left[\Phi_{h,f\to s}^{n} + \Phi_{h,f\to s}^{n+1} + \left(\Phi_{m,f\to s}^{n} + \Phi_{m,f\to s}^{n+1}\right)(-\Delta H_r)\right] \quad (19)$$

As indicated in beginning of Section 2 in immersed boundary techniques several fluid–solid (mass, momentum, and heat) coupling schemes can be adopted. In our technique, we enforce the (Dirichlet) boundary conditions at the level of the discrete form of the field equations with the method described in more detail in Deen et al. (2012) and Deen and Kuipers (2013, 2014). The advantages of this method are its relative simple computer implementation and second-order accuracy. The second feature is beneficial when dealing with many-particle systems because it allows to use a relatively low number of grid points distributed over the particle diameter.

More complex boundary conditions including the treatment of, for instance, conjugate mass, heat transport, and surface reactions can be treated as well (see, respectively, Das et al., 2017 and Lu et al., 2018).

2.3 Results DNS

2.3.1 Flow and Heat Transfer in Stationary Particle Arrays

As an example to demonstrate the capabilities of the DNS approach, in this section we will consider the flow through a stationary random array of hundreds of particles accompanied by mass and heat transfer and exothermic chemical reaction at the surface of the particles (see Table 1 for the data used

Table 1 Data Used for Simulation of Mass Transfer Limited Exothermal Chemical Reaction at the Surface of a Stationary Array of Spheres

Computational grid	$200 \times 200 \times 600 = 24{,}000{,}000$	(–)
Grid size	0.0005	m
Time step	0.0001	s
Particle radius	0.0060	m
Initial sphere positions	Random particle configuration (663 spheres) $0.0 < x\text{-position} < 0.100$ $0.0 < y\text{-position} < 0.100$ $0.0 < z\text{-position} < 0.200$ bed voidage $\varepsilon = 0.7001$	(–) m m m
Fluid density	1.0	kg/m^3
Fluid viscosity	2.0×10^{-5}	kg/(m s)
Fluid diffusivity	2.0×10^{-5}	m^2/s
Fluid thermal conductivity	0.025	W/(m K)
Fluid heat capacity	1000.0	J/(kg K)
Fluid inlet temperature	293.0	K
Fluid initial temperature	293.0	K
Particle volumetric heat capacity	1000.0	J/m^3
Particle initial temperature	293.0	K
Reaction heat	−100,000	J/(mole)
Particle Reynolds number	180	(–)

in the simulation). Further details of the model can be found in Deen et al. (2012) and Deen and Kuipers (2013, 2014) as well as extensive verification and validation of the computational methods. The spheres were distributed in a random fashion over the computational domain (excluding the top section with $z > 0.20$ m) using a standard Monte Carlo procedure to produce a predefined voidage of approximately 0.70. For our calculations, we use $d_p/h = 24$, with h the uniform grid spacing. The incorporation of the empty top section was found to be essential to avoid problems due to recirculation over the outflow boundary at the chosen particle Reynolds numbers ($Re_p > 180$). For the simulation, a prescribed uniform velocity and a prescribed pressure were imposed at, respectively, the bottom and top boundaries. For the other boundaries the no-slip condition was imposed. At the bottom wall, the fluid species concentration and the fluid temperature were set at 1.0 mol/m^3 and 293.0 K, respectively. At all remaining walls zero concentration and temperature gradients (homogeneous Neumann boundary condition) were imposed.

For the other lateral boundaries the no-slip condition was imposed. At the bottom wall, the fluid the species concentration and the fluid temperature were set at 1.0 mol/m^3 and 293.0 K, respectively. For the species conservation equation and the thermal energy equation, respectively, zero concentration and temperature gradients (Neumann boundary condition) were imposed.

The thermal behavior of the system is complex and involves (i) transient heating of particles due to the exothermic chemical reaction at the particle surface; (ii) subsequent heat transfer from the heated particles to the fluid flowing through the interstitial void spaces; and (iii) heat transfer from the fluid to downstream particles. This sequence of steps in fact implies that temperature waves propagate through the domain where the heat produced by the exothermic chemical reaction temporarily stored in the particles is swept downstream by the fluid and transferred to relatively cold particles. In Fig. 4, the evolution of the temperature distributions in the system is shown for the vertical central plane (the particles are excluded from the figures) at $Re_p = 180$. From Fig. 4, it can be seen that particles with (nearly) the same axial coordinate experience different heating rates. This is due to substantial differences in the local fluid–particle mass and heat transfer coefficients caused by the existence of (local) preferential pathways for the fluid. From the DNS results detailed information can be obtained such as the (distribution of) the mass and heat transfer coefficients experienced

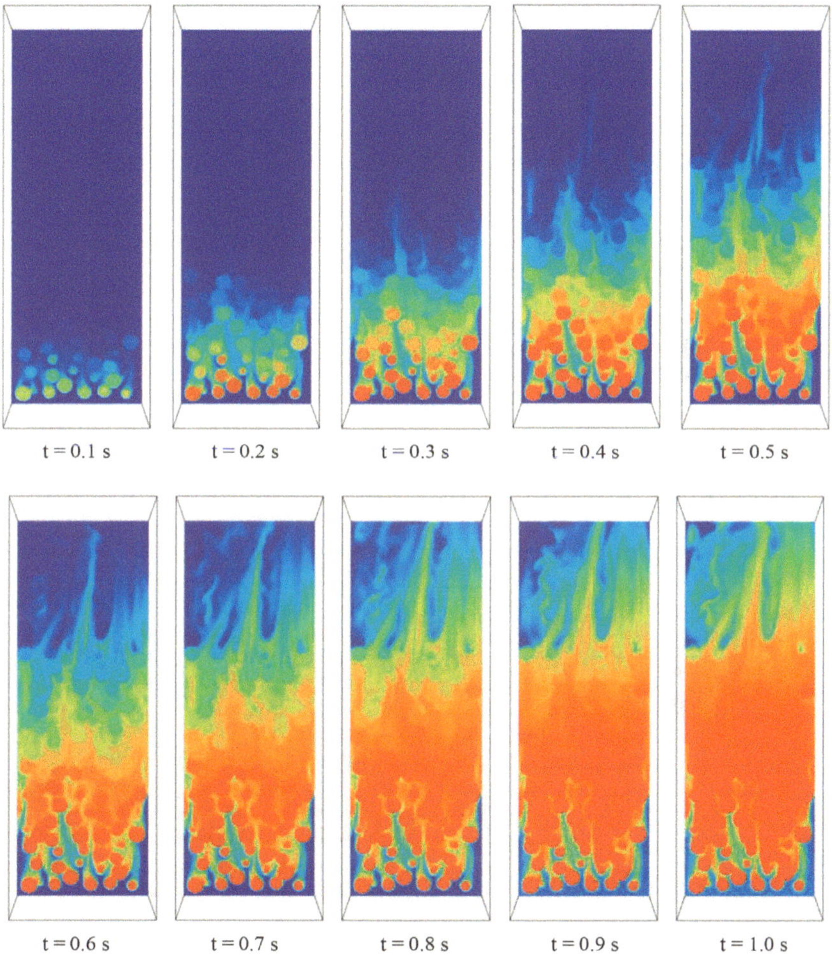

Fig. 4 Evolution of the temperature distribution for $Re_p = 180$ visualized for the central (vertical) *yz*-plane. The particles are not included in this figure (*blue*: temperature 293.0 K, *red*: temperature 393 K). *Reproduced with permission from Deen NG, Kuipers JAM: Direct numerical simulation of fluid flow and mass transfer in dense fluid–particle systems,* Ind Eng Chem Res *52:11266–11274, 2013.*

by individual particles. Here we define the local fluid–particle mass and heat transfer coefficients k_m and α_p by the following equations:

$$k_m = \frac{\Phi_{m,f \to s}}{4\pi R_p^2 \langle c_{A,f} \rangle} \tag{20}$$

$$\alpha_{\mathrm{p}} = \frac{\Phi_{\mathrm{h,f}\rightarrow \mathrm{s}}}{4\pi R_{\mathrm{p}}^2 \left(T_{\mathrm{p}} - \langle T_{\mathrm{f}} \rangle \right)} \tag{21}$$

where $\Phi_{\mathrm{m,f}\rightarrow \mathrm{s}}$ and $\Phi_{\mathrm{h,f}\rightarrow \mathrm{s}}$ represent the mass and heat transfer rates of the surrounding fluid to the particle defined by:

$$\Phi_{\mathrm{m,f}\rightarrow \mathrm{s}} = -\iint_{S_{\mathrm{p}}} (D_{\mathrm{A,f}} \nabla c_{\mathrm{A,f}} \cdot \overline{n}) \mathrm{d}S \tag{22}$$

$$\Phi_{\mathrm{h,f}\rightarrow \mathrm{s}} = -\iint_{S_{\mathrm{p}}} (\kappa_{\mathrm{f}} \nabla T_{\mathrm{f}} \cdot \overline{n}) \mathrm{d}S \tag{23}$$

Note that for the evaluation of the local driving force in Eqs. (20) and (21) the local (i.e., corresponding to the local z-value) cup-mixing concentration and cup-mixing temperature are used. These quantities are defined as:

$$\langle c_{\mathrm{A,f}} \rangle = \frac{\iint_{S_{\mathrm{f}}} u_z(x, y, z) c_{\mathrm{A,f}}(x, y, z) \mathrm{d}x\mathrm{d}y}{\iint_{S_{\mathrm{f}}} u_z(x, y, z) \mathrm{d}x\mathrm{d}y} \tag{24}$$

$$\langle T_{\mathrm{f}} \rangle = \frac{\iint_{S_{\mathrm{f}}} u_z(x, y, z) T_{\mathrm{f}}(x, y, z) \mathrm{d}x\mathrm{d}y}{\iint_{S_{\mathrm{f}}} u_z(x, y, z) \mathrm{d}x\mathrm{d}y} \tag{25}$$

In Table 2, the average mass and heat transport coefficients that are computed using the settings listed in Table 1 are compared (at $t = 1.0$ s) with the corresponding values obtained from the Gunn correlation:

$$\begin{aligned} Nu_{\mathrm{p}} = \frac{\alpha_{\mathrm{p}} d_{\mathrm{p}}}{\kappa_{\mathrm{f}}} &= \left(7 - 10\varepsilon_{\mathrm{f}} + 5\varepsilon_{\mathrm{f}}^2\right)\left(1 + 0.7\left(Re_{\mathrm{p}}\right)^{0.2}(Pr)^{1/3}\right) \\ &+ \left(1.33 - 2.40\varepsilon_{\mathrm{f}} + 1.20\varepsilon_{\mathrm{f}}^2\right)\left(Re_{\mathrm{p}}\right)^{0.7}(Pr)^{1/3} \end{aligned} \tag{26}$$

Table 2 Average Values of the Flow-Averaged Mass and Heat Transfer Coefficients Computed From the DNS Data in Comparison With the Mass and Heat Transfer Coefficients Computed From the Gunn (1978) Correlation Using the Data Given in Table 1

Re_{p}	α_{p} (W/(m^2 K)) DNS	α_{p} (W/(m^2 K)) Gunn (1978)	k_{m} (m/s) DNS	k_{m} (m/s) Gunn (1978)
120	25.23	26.87	0.0219	0.0228
180	30.28	31.91	0.0263	0.0272
240	34.33	36.36	0.0298	0.0310

where Re_{p} and Pr, respectively, represent the particle Reynolds number (based on the superficial velocity U_0) and the Prandtl number:

$$Re_{\mathrm{p}} = \frac{\varepsilon_{\mathrm{f}}\rho_{\mathrm{f}}\left|\overline{u} - \overline{w}_{\mathrm{p}}\right| d_{\mathrm{p}}}{\mu_{\mathrm{f}}} \quad Pr = \frac{\mu_{\mathrm{f}} C_{\mathrm{p,f}}}{\kappa_{\mathrm{f}}} \tag{27}$$

For the evaluation of the fluid–particle heat transfer coefficients particles which have not yet attained their final temperatures are excluded. As can be seen from this table, the agreement between the DNS data and the values obtained from the Gunn correlation is very good.

In Fig. 5, a comparison is presented between the DNS data and results obtained from a 1D heterogeneous model which uses the empirical correlation of Gunn (1978) to evaluate the average value of the mass and heat transfer coefficients.

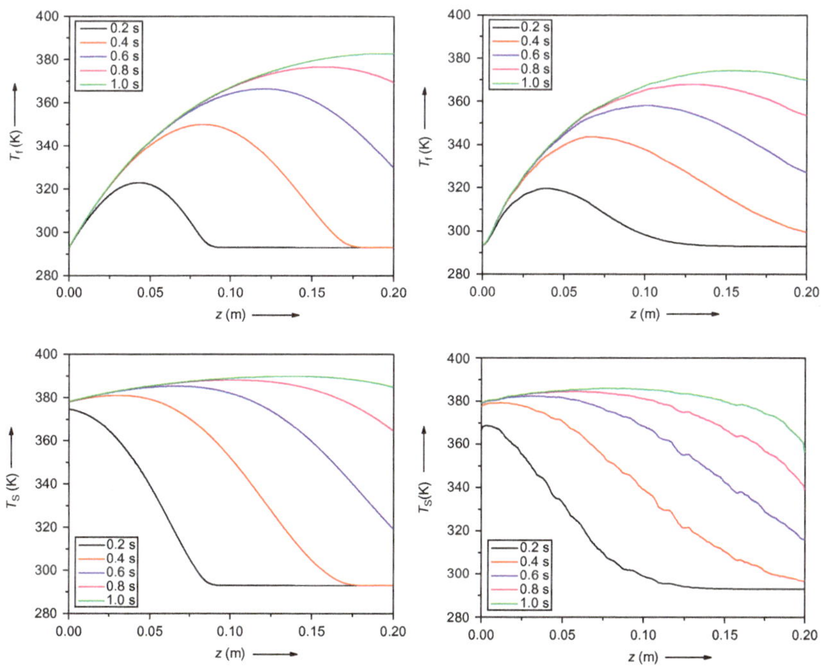

Fig. 5 Vertical profiles of the flow-averaged fluid temperature (*top*) and the laterally averaged solid (*bottom*) temperature obtained from a simple one-dimensional (1D) heterogeneous model (*left*) and DNS (*right*) at $Re_{\mathrm{p}} = 180$. *Reproduced with permission from Deen NG, Kuipers JAM: Direct numerical simulation of fluid flow accompanied by coupled mass and heat transfer in dense fluid–particle systems,* Chem Eng Sci *116:645–656, 2014.*

The overall features of the transient temperature profiles agree very well but small differences can be observed. First of all, both the maximum temperature and the temperature gradient beyond the maximum obtained from our DNS are lower compared to the results from the 1D heterogeneous model based on plug flow. This difference is caused by the nonuniform velocity profile due to the presence of the particles. Effectively this introduces (minor) axial dispersion of heat and consequently a lower maximum temperature and axial gradient of the temperature.

2.3.2 Fluidization of Spheres in a Pseudo 2D Bed

As a second example, the fluidization of 1296 particles (see Table 3 for the data used in the simulation) will be presented where initially the particles are at rest and distributed over the bottom half of the domain in a lattice with uniform voidage of 0.65. In this case the bed thickness in the depth direction slightly exceeds the particle diameter. The height of the particle free zone is the same as the initial bed height, allowing for expansion of the particle suspension. No-slip boundaries were taken at the walls confining the bed in all the lateral directions whereas a prescribed pressure boundary was taken at the

Table 3 Data Used for the Pseudo 2D Fluidized Bed of Spheres

Computational grid	$400 \times 12 \times 800 = 3{,}840{,}000$	(−)
Grid size	0.0005	m
Time step	0.0002	s
Particle radius	0.0025	m
Initial sphere positions	Uniform lattice of $(36 \times 36) = 1296$ Spheres in bottom half of the domain	(−)
Fluid density	1216	kg/m^3
Fluid viscosity	0.100	kg/(m s)
Fluid thermal conductivity	1.0	W/(m K)
Fluid heat capacity	10	J/(kg K)
Superficial liquid velocity	0.08	m/s
Solid density	8000	kg/m^3
Collision parameters e_n, e_t, μ_t	0.90, 0.30, 0.10	(−)

top of the domain. The minimum fluidization velocity for the fluid–particle combination was estimated from the Ergun equation as 1.17 cm/s whereas the terminal velocity of the particle amounted 39.6 cm/s. For the simulation of this thin pseudo 2D fluidized bed, the superficial velocity was set at 8 cm/s.

In Fig. 6, the evolution of the bed particle configuration is shown as a function of time. It can be seen that initially the bed expands uniformly, however near the left and right walls of the bed (due to the no-slip condition) the particle configuration is disturbed producing eventually small cavities which travel near the walls in the vertical direction. In the central part of the bed also disturbances in the particle configuration develop and several cavities start to form and propagate in the vertical direction leading to breakage of the symmetry and consequently a "chaotic" movement and mixing of the particle bed results. Because the superficial velocity u_0 is well above, the minimum fluidization velocity significant expansion of the particle bed occurs. Once the bed expansion is stable the total force exerted by the fluid on particle bed in the vertical direction equals the total weight of the particles (difference less than 1%) however, this balancing of forces does not hold at the level of individual particles. Fig. 7 shows the velocity field of the fluidizing agent in the central plane for a small subdomain of the bed at $t = 2.0$ s revealing a highly nonuniform flow field. The detailed information obtained from this type of simulation is very useful to develop and assess the (drag and heat transfer) closures adopted in models which do not resolve all the details of the flow of the fluidizing agent. The widely used CFD–DEM is a well-known example where typically the computational cells are much larger than the size of an individual particle (Van der Hoef et al., 2008) necessitating specification of closures for the fluid–particle interactions. The average fluid–particle heat transfer coefficient equals 789.0 W/(m^2 K).

From the bed expansion in the bottom zone of the fluidized bed a (nearly constant) average voidage of 0.80 can be calculated which corresponds to an average fluid–particle heat transfer coefficient of 970.0 W/(m^2 K) according to the empirical correlation proposed by Gunn (1978).

Besides validating existing closure relations the DNS approach have also been used to estimate closure relation for fluid–particle drag (Das et al., 2016; Mehrabadi et al., 2016; Tang et al., 2015) and fluid–solid heat transfer (Das et al., 2016; Deen et al., 2014). These improved closure relations can next be used in coarser models such as DEM and TFM.

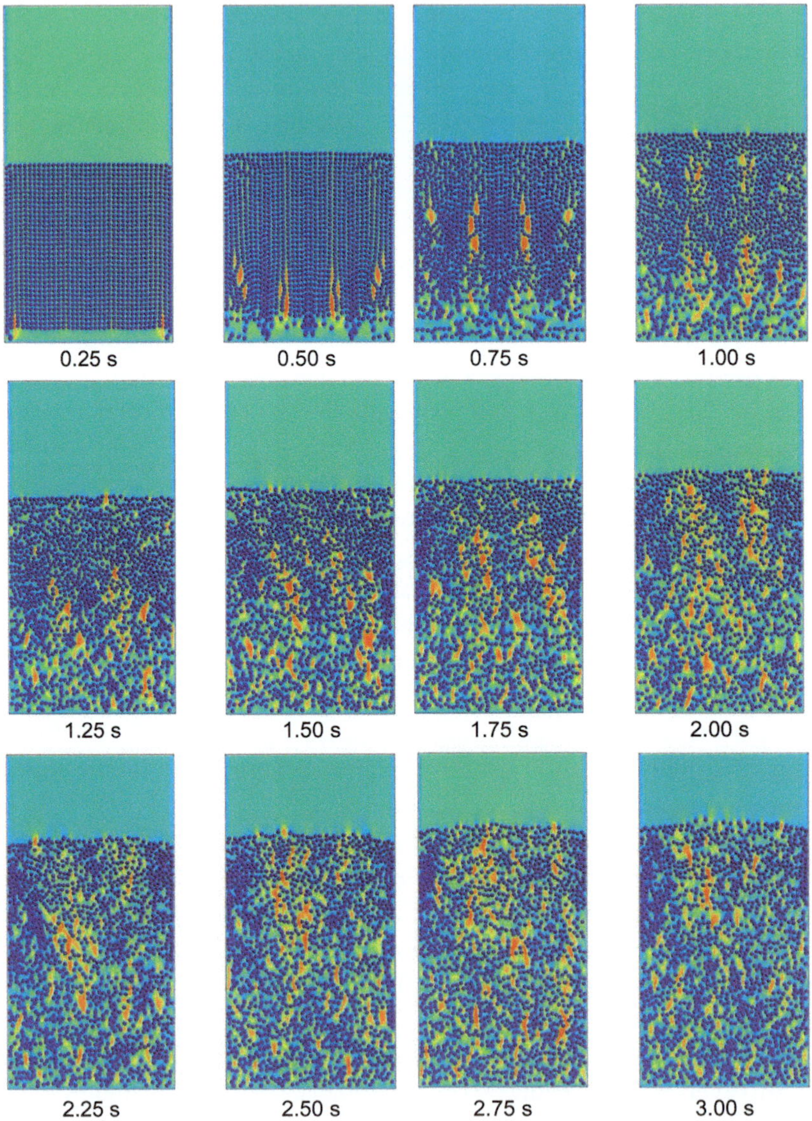

Fig. 6 Evolution of bed structure as a function of time for fluidization of 1296 spheres at superficial velocity of 8 cm/s. Incipient fluidization velocity 1.17 cm/s; terminal velocity 39.6 cm/s. *Reproduced with permission from Deen NG, Kriebitzsch SHL, van der Hoef MA, Kuipers JAM: Direct numerical simulation of flow and heat transfer in dense fluid–particle systems,* Chem Eng Sci *81:329–344, 2012.*

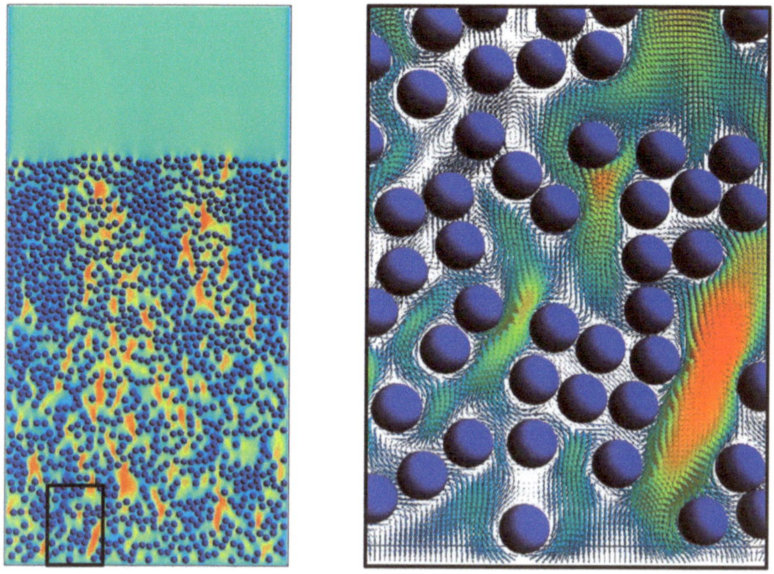

Fig. 7 Particle configuration (*left*) and velocity map (*right*) at the central plane for a small subdomain of the bed at $t = 2.00$ s. Incipient fluidization velocity 1.17 cm/s; terminal velocity 39.6 cm/s. *Reproduced with permission from Deen NG, Kriebitzsch SHL, van der Hoef MA, Kuipers JAM: Direct numerical simulation of flow and heat transfer in dense fluid–particle systems,* Chem Eng Sci *81:329–344, 2012.*

3. DEM

In the DEM, the volume-averaged Navier–Stokes equations are solved for the continuous phase whereas the Newtonian equations of motion are solved for each particle while accounting for particle–particle and particle–wall collisions. This type of model can be classified as an unresolved Euler–Lagrange (EL) type of model since typically (unlike the DNS or PRS approach) the computational grid is (much) larger than the size of individual particles and consequently closures for fluid–particle interaction need to be specified either on basis of empirical data or parameterization of DNS data. Both the hard sphere and soft sphere approach have been successfully used to simulate dense gas-fluidized beds.

The hard sphere model assumes binary quasi-instantaneous dissipative collisions and consequently this model is not suited to study granular flow with static regions where multiple particle–particle contacts prevail. For such systems, the soft sphere model is a better choice because it can efficiently handle multiple particle–particle interactions where the pair-wise interactions are described on basis of a mechanical model comprising a spring, a friction slider, and a dashpot. The SIMPLE algorithm (Patankar and Spalding, 1972) is used to solve the conservation equations for the continuous phase while the equations of motion for the particulate phase are solved with standard ODE solvers. Efficient algorithms from the field of molecular dynamics (MD) are used for the efficient processing of the large number of (dissipative) collisions. DEM has received a lot of attention in recent years to study complex particulate flows, for further details we refer to the review by Deen et al. (2007).

3.1 Governing Equations DEM

The system of interest requires the simultaneous solution of the gas phase and solid (particle) phase equations. For the model presented in this chapter, we make use of the following main assumptions:

- The fluid phase (gas) is compressible and obeys the ideal gas law.
- The thermophysical properties of both phases are constant.
- The solid phase consists of spherical catalyst particles.
- Intraparticle temperature gradients are negligible.

3.1.1 Fluid Phase Equations

The governing conservation equations for dense fluid–solid flows are given by the following expressions:

$$\frac{\partial}{\partial t}(\varepsilon_{\mathrm{f}}\rho_{\mathrm{f}}) + (\nabla\cdot\varepsilon_{\mathrm{f}}\rho_{\mathrm{f}}\overline{u}) = 0 \tag{28}$$

$$\frac{\partial}{\partial t}(\varepsilon_{\mathrm{f}}\rho_{\mathrm{f}}\overline{u}) + (\nabla\cdot\varepsilon_{\mathrm{f}}\rho_{\mathrm{f}}\overline{\overline{u}\overline{u}}) = -\varepsilon_{\mathrm{f}}\nabla p - (\nabla\cdot\varepsilon_{\mathrm{f}}\tau_{\mathrm{f}}) - \overline{f}_{\mathrm{f}\rightarrow\mathrm{p}} + \varepsilon_{\mathrm{f}}\rho_{\mathrm{f}}\overline{g} \tag{29}$$

$$\rho_{\mathrm{f}}C_{\mathrm{p,f}}\left[\frac{\partial}{\partial t}(\varepsilon_{\mathrm{f}}T_{\mathrm{f}}) + (\nabla\cdot\varepsilon_{\mathrm{f}}\overline{u}T_{\mathrm{f}})\right] = -\left(\nabla\cdot\overline{\Phi}_{\mathrm{f}}\right) - h_{\mathrm{f}\rightarrow\mathrm{p}} \tag{30}$$

For the fluid phase stress tensor the general Newtonian form is assumed, given by (Bird et al., 1960):

$$\tau_{\mathrm{f}} = -\left(\lambda_{\mathrm{f}} - \frac{2}{3}\mu_{\mathrm{f}}\right)(\nabla\cdot\overline{u})I - \mu_{\mathrm{f}}\left[(\nabla\overline{u}) + (\nabla\overline{u})^{T}\right] \tag{31}$$

where μ_{f} and λ_{f} are the shear and bulk viscosity of the fluid. For the heat flux $\overline{\Phi}_{\mathrm{f}}$ Fourier's law is assumed:

$$\overline{\Phi}_{\mathrm{f}} = -\kappa_{\mathrm{e,f}}\nabla T_{\mathrm{f}} \tag{32}$$

where $\kappa_{e,f}$ represents the effective thermal conductivity of the fluid for which we use the following expression:

$$\kappa_{e,f} = \left(1 - \sqrt{(1-\varepsilon_f)}\right)\kappa_f \tag{33}$$

where κ_f represents the thermal conductivity of the fluid. The source terms $\overline{f}_{f\to p}$ and $h_{f\to p}$ account, respectively, for the two-way coupling of momentum and heat which are obtained from the Lagrangian part of the model which will be discussed later.

3.1.2 Particle phase equations

The motion of the suspended solid particles is given by Newton's second law given by the following expression which accounts for the action of gravity, far-field pressure, and drag.

$$\frac{\mathrm{d}}{\mathrm{d}t}\left(m_p \overline{w}_p\right) = \overline{F}_g + \overline{F}_p + \overline{F}_d = m_p \overline{g} - V_p \nabla p + \frac{V_p \beta_p}{(1-\varepsilon_f)}\left(\overline{u} - \overline{w}_p\right) \tag{34}$$

Since the size of the particles typically is (much) smaller than the size of the computational grid required for capturing the bubble dynamics, a closure equation is required for the effective drag (friction) acting on the particles. Following Hoomans et al. (1996) the friction between the continuous liquid phase and the suspended solid particles is given by the Ergun equation in the dense regime ($\varepsilon_f < 0.8$) and the Wen and Yu (1966) equation in the dilute regime ($\varepsilon_f > 0.8$), respectively, given by

$$(\varepsilon_f < 0.8)\beta_p = 150\frac{(1-\varepsilon_f)^2}{\varepsilon_f}\frac{\mu_f}{(d_p)^2} + 1.75(1-\varepsilon_f)\frac{\rho_f}{d_p}\left|\overline{u} - \overline{w}_p\right| \tag{35}$$

and

$$(\varepsilon_f > 0.8)\beta_p = \frac{3}{4}C_d\frac{\varepsilon_f(1-\varepsilon_f)}{d_p}\rho_f\left|\overline{u} - \overline{w}_p\right|(\varepsilon_f)^{-2.65} \tag{36}$$

with the single particle drag coefficient for spheres:

$$C_d = \begin{cases} \dfrac{24}{Re_p}\left(1 + 0.15{Re_p}^{0.687}\right) & Re_p < 1000 \\ 0.44 & Re_p > 1000 \end{cases} \tag{37}$$

Alternatively drag closure equations obtained from LB models can be used for the effective drag closure (Van der Hoef et al., 2008, 2004).

For the (possible) nonideal particle–particle and particle–wall collisions, we use a three-parameter model accounting for normal and tangential restitution and friction. The thermal energy equation for an individual particle is given by:

$$m_{\mathrm{p}} C_{\mathrm{p,p}} \frac{\mathrm{d} T_{\mathrm{p}}}{\mathrm{d} t} = -4\pi R_{\mathrm{p}}^{2} \alpha_{\mathrm{p}} \left(T_{\mathrm{p}} - T_{\mathrm{f}} \right) \tag{38}$$

where $C_{\mathrm{p,p}}$ represents the particle heat capacity and α_{p} is the fluid–particle heat transfer coefficient given by the empirical equation proposed by Gunn (Eq. 26).

3.2 Numerical Solution Method DEM

First we perform the time discretization of the momentum equation to obtain the following expression:

$$(\varepsilon_{\mathrm{f}} \rho_{\mathrm{f}} \overline{u})^{n+1} = \overline{S}_{\mathrm{m}}^{n} - \Delta t \left[(\varepsilon_{\mathrm{f}} \nabla p)^{n+1} + \left(\overline{f}_{\mathrm{f} \to \mathrm{p}} \right)^{n+1} \right] \tag{39}$$

where n indicates the time index whereas the explicit terms have been collected in $\overline{S}_{\mathrm{m}}^{n}$ given by:

$$\overline{S}_{\mathrm{m}}^{n} = (\varepsilon_{\mathrm{f}} \rho_{\mathrm{f}} \overline{u})^{n} - \Delta t [(\nabla \bullet \varepsilon_{\mathrm{f}} \rho_{\mathrm{f}} \overline{u u})^{n} + (\nabla \cdot \varepsilon_{\mathrm{f}} \tau_{\mathrm{f}})^{n} + (\varepsilon_{\mathrm{f}} \rho_{\mathrm{f}})^{n} \overline{g}] \tag{40}$$

The viscous terms are treated with standard central second-order finite difference representations whereas the convective terms are computed with a second-order flux delimited Barton scheme (Centrella and Wilson (1984). A key element in the solution of the momentum equations and the thermal energy equation is the linear implicit treatment of the source terms $\overline{f}_{\mathrm{f}\to\mathrm{p}}$ and $h_{\mathrm{f}\to\mathrm{p}}$ which account for the two-way coupling caused by the presence of the particles. These terms are now written in, respectively, a part which is independent of the fluid velocity and fluid temperature and a part which, respectively, is proportional to the fluid velocity and the fluid temperature:

$$\overline{f}_{\mathrm{f}\to\mathrm{p}}^{n+1} = \overline{\alpha}_{\mathrm{f}\to\mathrm{p}}^{n}(\overline{r}) + \beta_{\mathrm{f}\to\mathrm{p}}^{n}(\overline{r}) \overline{u}(\overline{r})^{n+1} = \frac{1}{V_{\mathrm{cell}}} \sum_{p=1}^{p=np} \overline{F}_{\mathrm{d}}(\overline{r}_p) D(\overline{r} - \overline{r}_p) \tag{41}$$

$$h_{\mathrm{f}\to\mathrm{p}}^{n+1} = \phi_{\mathrm{f}\to\mathrm{p}}^{n}(\overline{r}) + \varphi_{\mathrm{f}\to\mathrm{p}}^{n}(\overline{r}) T_{\mathrm{f}}(\overline{r})^{n+1} = \frac{1}{V_{\mathrm{cell}}} \sum_{p=1}^{p=np} \Phi_{\mathrm{h}}(\overline{r}_p) D(\overline{r} - \overline{r}_p) \tag{42}$$

where $\overline{\alpha}_{\mathrm{f}\to\mathrm{p}}^{n}$, $\beta_{\mathrm{f}\to\mathrm{p}}^{n}$, $\phi_{\mathrm{f}\to\mathrm{p}}^{n}$, and $\varphi_{\mathrm{f}\to\mathrm{p}}^{n}$ are obtained from the Lagrangian part of the model for the particles described hereafter. In Eqs. (41) and (42), the summation term accounts for the contributions of all particles p which have their center of mass (given by their position vectors $\bar{r}_p$) within a certain computational cell with volume $V_{\text{cell}} = \Delta x \Delta y \Delta z$. In these expressions, $D(\bar{r})$ represents a distribution function which is decomposed in separate spatial contributions according to the following expression:

$$D(\bar{r}) = D(x, y, z) = D_x(x) D_y(y) D_z(z) \tag{43}$$

For the distribution function $D(\bar{r})$, we typically use volume weighing whereas for the evaluation of the source terms in the momentum equation, care should be taken to account for the staggered computational grid. The sum of the interphase forces is obtained from the Lagrangian part of the model. This particular (linear implicit) treatment of the source terms greatly enhances the numerical stability of the model, especially at high particle volume fraction. Combination of Eqs. (39) and (41) leads to the following expression for the momentum density at the advanced time level:

$$(\varepsilon_f \rho_f \bar{u})^{n+1} = \overline{S}^n_{\text{m}} - \Delta t \left[\left(\varepsilon_f^{n+1} \nabla p^{n+1} \right) + \overline{\alpha}^n_{\text{f} \to \text{p}} + \beta^n_{\text{f} \to \text{p}} \bar{u}^{n+1} \right] \tag{44}$$

or after collecting terms:

$$\left[(\varepsilon_f \rho_f)^{n+1} + \Delta t \beta^n_{\text{f} \to \text{p}} \right] \bar{u}^{n+1} = \overline{S}^n_{\text{m}} - \Delta t \left[\left(\varepsilon_f^{n+1} \nabla p^{n+1} \right) + \overline{\alpha}^n_{\text{f} \to \text{p}} \right] \tag{45}$$

Discretization of the time derivative in the continuity equation leads to the following expression which defines the mass residual D_f^{n+1}:

$$(\varepsilon_f \rho_f)^{n+1} - (\varepsilon_f \rho_f)^n + \Delta t \left[\nabla \cdot (\varepsilon_f \rho_f \bar{u})^{n+1} \right] = D_f^{n+1} \tag{46}$$

This mass residual for a certain computational cell (i,j,k) is considered to depend on the pressure in the same computational cell and the pressures of the six surrounding computational cells:

$$\begin{aligned} (D_f)^{n+1}_{i,j,k} &= (D_f)^{n+1}_{i,j,k} \left(p^{n+1}_{i-1,j,k}, p^{n+1}_{i+1,j,k}, p^{n+1}_{i,j-1,k}, p^{n+1}_{i,j+1,k}, p^{n+1}_{i,j,k-1}, p^{n+1}_{i,j,k+1}, p^{n+1}_{i,j,k} \right) \\ &\equiv 0 \end{aligned} \tag{47}$$

which after linearization leads to the following pressure correction equation in vector notation:

$$J \delta \bar{p}^{n+1} = -\overline{D}_f^* \tag{48}$$

where J represents the assembled Jacobi matrix, $\delta\overline{p}^{n+1}$ the vector of (unknown) pressure corrections, and $\overline{D}_f^*$ the vector of known mass residuals computed from the most recently available hydrodynamic field variables. The elements of J can be computed analytically from the definition equation for the mass residual (using central finite difference approximations for the convection terms) and the momentum equation. The resulting (heptadiagonal) sparse matrix equation is again solved with a very efficient ICGM. For each time step only a few (3–4) implicit iterations are required to attain convergence. Following the solution of the continuity and momentum equations, the thermal energy equation of the fluid is solved.

$$\rho_\mathrm{f} C_{\mathrm{p,f}}(\varepsilon_\mathrm{f} T_\mathrm{f})^{n+1} = S_\mathrm{h}^n + \Delta t\left[(\nabla \cdot \kappa_{\mathrm{f,e}} \nabla T_\mathrm{f})^{n+1} - (h_{\mathrm{f}\rightarrow\mathrm{p}})^{n+1}\right] \tag{49}$$

where n indicates the time index whereas S_h^n represents the explicit term given by:

$$S_\mathrm{h}^n = \rho_\mathrm{f} C_{\mathrm{p,f}}(\varepsilon_\mathrm{f} T_\mathrm{f})^n - \Delta t \rho_\mathrm{f} C_{\mathrm{p,f}}(\nabla \bullet \varepsilon_\mathrm{f} T_\mathrm{f} \overline{u})^n \tag{50}$$

Combination of Eqs. (42) and (49) leads to the following expression for the thermal energy equation:

$$\left[\rho_\mathrm{f} C_{\mathrm{p,f}} \varepsilon_\mathrm{f}^{n+1} + \Delta t \varphi_{\mathrm{f}\rightarrow\mathrm{p}}^n\right] T_\mathrm{f}^{n+1} = S_\mathrm{h}^n + \Delta t\left[(\nabla \cdot \kappa_{\mathrm{e,f}} \nabla T_\mathrm{f})^{n+1} - \phi_{\mathrm{f}\rightarrow\mathrm{p}}^n\right] \tag{51}$$

For the spatial discretization of the convection and conduction terms in, respectively, (50) and (51) exactly the same schemes are used as for the solution of the momentum equations.

The source terms appearing in the Newtonian equations of motion are treated as known (explicit) terms and therefore the integration of these equations can be conducted in principle with any integration technique for ordinary differential equations. For the simulations reported in this chapter, we have used a simple first-order explicit scheme producing translational and rotational particle velocities at the new time level. Once the particle velocities are obtained a collision model (either the hard sphere or soft sphere model) is invoked to account for the particle–particle and particle–wall collisions. The closure of this collision model involves three micromechanical parameters: the coefficients for normal and tangential restitution and the tangential friction coefficient, which in principle can be obtained from impact experiments (see Deen et al., 2007). The particle phase thermal energy equation is solved via a first-order linear implicit scheme to obtain the particle temperature at the new time level.

3.3 Results DEM

3.3.1 Gas-Fluidized Beds With Heat Production

Relatively few studies have been reported on the (simultaneous) mass and heat transfer in dense fluid–particle systems. In this section, we will study a gas-fluidized bed with heat production in the particulate phase. This type of system is very often encountered in large-scale processes using dense gas-fluidized beds such as combustion, gasification, partial oxidation of hydrocarbons, and gas phase polymerization for the production of poly-olefines. In the latter process, the monomer (ethylene or propylene) is contacted with a prepolymerized particles containing metallocene type of catalysts leading to very rapid growth of the particles with liberation of a large amount of heat due to the strongly exothermic polymerization reactions.

To avoid the full complexity of catalytic olefin polymerization in the gas phase Li et al. (2017) used CO_2 adsorption on Zeolite 13X particles (1.8–2.0 mm, Geldart D type) as a model system to mimic the heat effect due to the exothermic polymerization reaction. All key aspects of this process (kinetics, equilibrium, and heat effect) were studied separately using thermogravimetric analysis (TGA) and simultaneous thermal analysis (STA). A combined infrared/visual camera technique generated detailed data on the hydrodynamic and thermal behavior of a pseudo 2D bed equipped with a sapphire front glass to minimize the absorption of the emitted thermal radiation. They also compared the evolution of the spatial distribution of the measured particle temperature with CFD–DEM simulation results.

In Fig. 8, we show the comparison of the particulate phase temperature obtained from the experiment (top) with the results obtained from CFD–DEM simulations. From this figure, we can see that the temperature of the particles rises rapidly due to the heat production associated with the CO_2 adsorption. Gradually the adsorption equilibrium is obtained and due to the vanishing heat production and concurrent heat losses to the environment the bed gradually cools down. As evident from Fig. 6, the bed temperature is not uniform due to the fact that (among others) the fluid–particle heat transfer coefficient is a distributed property as shown by Deen et al. (2012) on basis of detailed DNS simulations for dense random arrays of particles. This is due to the fact that (even) in random dense arrays there exist locally preferred pathways of the fluid leading to locally different flow conditions and hence heat transfer rates.

For the CFD–DEM simulations, the Gunn (1978) correlation was used to evaluate the fluid–particle heat transfer coefficient. Deen and Kuipers (2014) showed that the fluid–particle mass and heat transfer coefficients

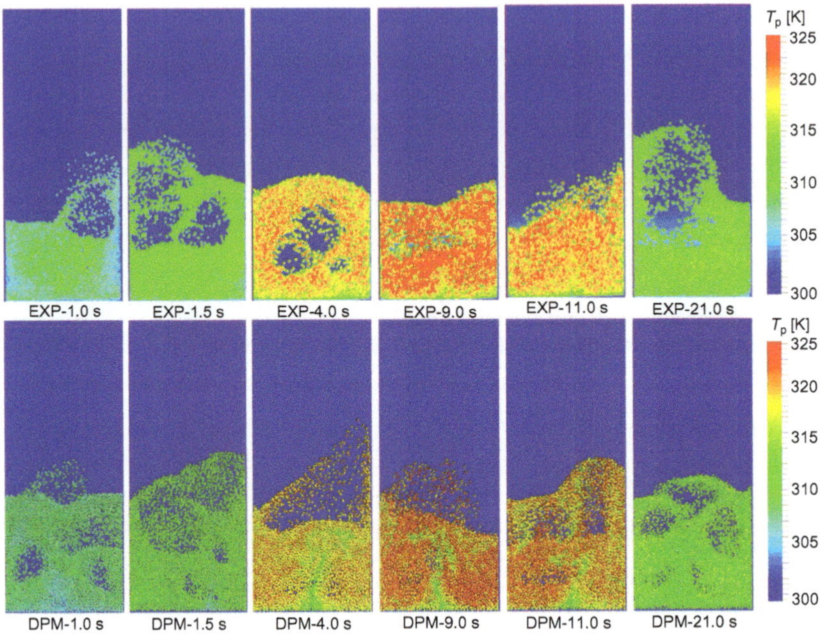

Fig. 8 Comparison of computed evolution of the particulate phase temperature obtained from experiment (*top*) using infrared (IR) imaging and CFD–DEM (DPM) simulation (*bottom*). Heat liberation is caused by the CO_2 adsorption on Zeolite 13X particles (1.8–2.0 mm diameter, Geldart D particles). *Reproduced with permission from Li Z, Janssen TCE, Buist KA, Deen NG, van Sint Annaland M, Kuipers JAM: Experimental and simulation study of heat transfer in fluidized beds with heat production,* Chem Eng J *317:242–257, 2017.*

for dense random arrays of particles obtained from DNS simulations agree very well with values obtained from the Gunn correlation. In Fig. 9, the distribution of the particle temperature obtained from the CFD–DEM simulations is compared with the experimental data in terms of the PDF and reveals a very good quantitative agreement between simulation and experiment.

3.3.2 Hydrodynamics of a Pseudo 2D Riser

Riser flows have been studied extensively by using both experimental and computational approaches. Carlos Varas et al. (2017a, 2017b) studied the hydrodynamics of a pseudo 2D riser using a combined digital image analysis (DIA) and particle image velocimetry (PIV). The advantage of the pseudo 2D geometry used by Carlos Varas et al. (2017a, 2017b) is given by the fact that direct visual observation of the heterogeneous flow structures

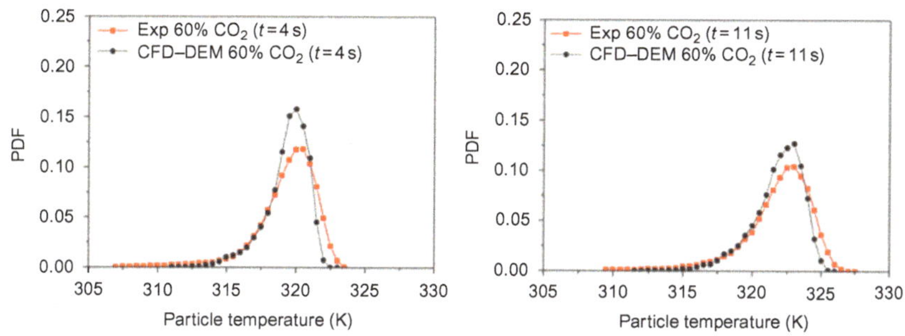

Fig. 9 Comparison of particulate phase temperature PDF obtained from experiment (*red*) using infrared and CFD–DEM simulation (*black*). Heat liberation is caused by the CO_2 adsorption on Zeolite 13X particles (1.8–2.0 mm diameter, Geldart D particles). *Reproduced with permission from Li Z, Janssen TCE, Buist KA, Deen NG, van Sint Annaland M, Kuipers JAM: Experimental and simulation study of heat transfer in fluidized beds with heat production,* Chem Eng J *317:242–257, 2017.*

(clusters or strands) is possible enabling detailed characterization and quantification using advanced DIA techniques (see Carlos Varas et al., 2016). Moreover, since the distance between the front and back plates is rather small, the flow is laminar and consequently uncertainties due to turbulence modelling can be circumvented. In Fig. 10, their experimental setup is shown revealing the presence of heterogeneous flow structures (cluster or strands) descending along the side walls of the riser. They also performed extensive and detailed CFD–DEM simulations and found that the main characteristics of the experimentally observed riser flow could be retrieved from these simulations.

In Fig. 11, both experimental and simulated lateral profiles of the time-averaged solids volume fraction at several heights in the riser are presented from this figure, it can be seen that at relatively low superficial gas velocities a clear lateral segregation occurs which diminishes at higher velocities. The agreement between the experimental and CFD–DEM simulation results is reasonable.

Carlos Varas et al. (2017b) also studied the impact of particle clustering on the gas–solid contact efficiency using detailed CFD–DEM simulations. From their study, it became clear that the gas phase partially circumvents descending dense particle clusters leading to lower gas–solid contact efficiencies. This by-passing induced by heterogeneous flow structures generally leads to lower performance of riser flow reactors as encountered in catalytic cracking processes.

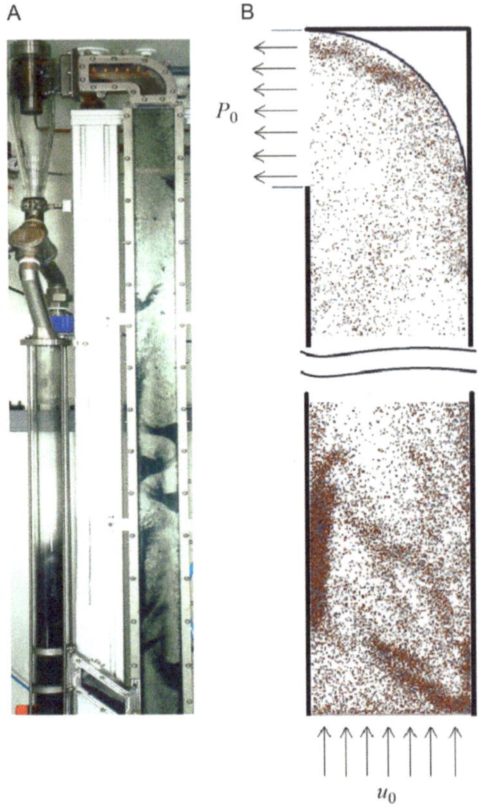

Fig. 10 Pseudo 2D riser (A) showing heterogeneous flow structures descending along the side walls of the riser. In addition, a snapshot (B) of a CFD–DEM simulation is included showing as well-heterogeneous flow structures. *Reproduced with permission from Carlos Varas AE, Peters EAJF, Kuipers JAM: CFD–DEM simulations and experimental validation of clustering phenomena and riser hydrodynamics,* Chem Eng Sci *169:246–258, 2017a.*

4. TFM

The TFM is based on the concept of interpenetrating continuous media and employs separate conservation equations for mass, momentum, and heat for each phase. As in the DEM the size of individual particles is much smaller that the size of the computational grid and consequently again closures for fluid–particle interaction need to be defined. Moreover closures for particle–particle interaction are required which in modern continuum models are based on the KTGF. The particle collision models employed in the TFM are typically accounting only for normal restitution. The extension of these

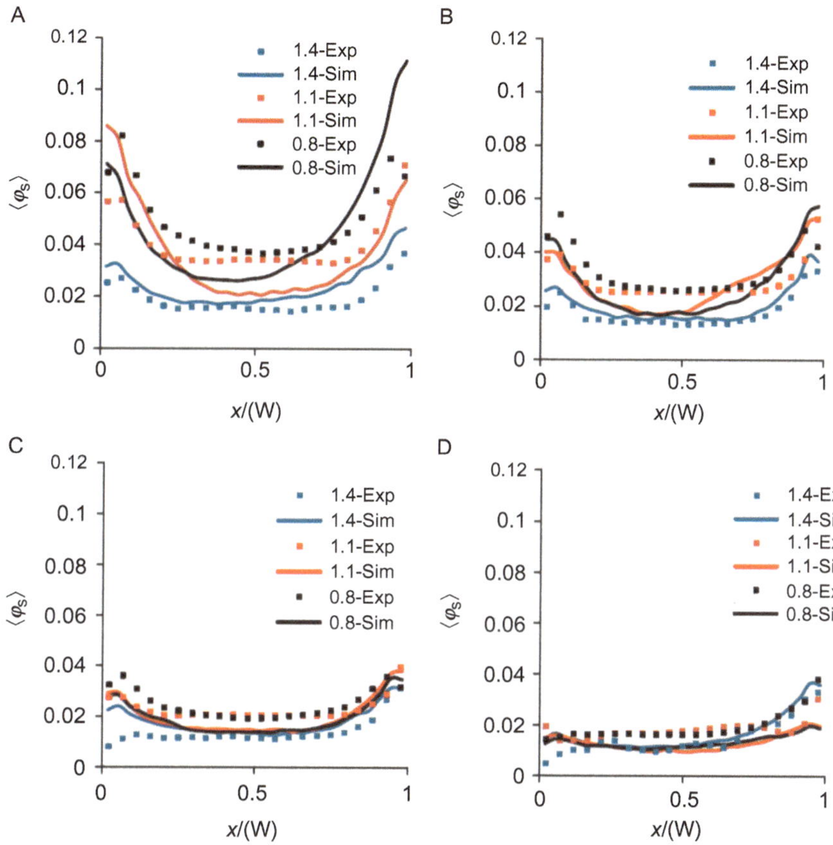

Fig. 11 Experimental and simulated lateral profiles of the time-averaged solids volume fraction at several heights in the riser for several superficial gas velocities. (A) $u_o = 5.55$ m/s, (B) $u_o = 5.95$ m/s, (C) $u_o = 6.35$ m/s, and (D) $u_o = 6.74$ m/s. *Reproduced with permission from Carlos Varas AE, Peters EAJF, Kuipers JAM: CFD–DEM simulations and experimental validation of clustering phenomena and riser hydrodynamics,* Chem Eng Sci *169:246–258, 2017a.*

models to include more realistic particle collision models accounting for particle friction as well as tangential restitution is quite challenging and has recently been reported by Yang et al. (2016a, 2016b) together with the appropriate boundary conditions at confining solid walls. Special algorithms are required to solve the governing equations which basically extend the well-known SIMPLE algorithm (Patankar and Spalding, 1972) to the so-called p–ε algorithm to enable the stable integration of the governing equations with a (substantially) bigger time step (Verma et al., 2013).

The organization of this section is as follows: first the governing and constitutive equations will be presented (Section 4.1) followed by a brief description of the numerical solution method (Section 4.2). In Section 4.3, the main results will be presented.

4.1 Governing Equations TFM

In the TFM, the concept of interpenetrating continuous media is used leading to separate conservation equations for mass, momentum, and heat for each phase (see Table 4). The main assumptions are:

- The fluid phase (gas) is compressible and obeys the ideal gas law.
- The thermophysical properties of both phases are constant.
- The solid phase can be modeled as a continuum and can be well represented by a mass density, velocity, and granular temperature field.
- Closure relations exist for intraparticle interactions (solids stress and pressure) and solid–fluid interaction (drag, transfer coefficients).

In addition to the governing equations, additional equations are required to close the set of equations. These closure equations are given in Table 5 and include equations for the interphase exchange of mass, momentum, and heat as well as closure equations for the stress tensors, fluctuating kinetic energy flux, and the transport coefficients. In addition, a closure for the dissipation

Table 4 TFM Governing Equations in Vector Form Based on the KTGF

Fluid phase continuity equation:
$\frac{\partial}{\partial t}\left(\varepsilon_g\rho_g\right)+\nabla\cdot\left(\varepsilon_g\rho_g\overline{u}_g\right)=0$
Solid phase continuity equation:
$\frac{\partial}{\partial t}(\varepsilon_s\rho_s)+\nabla\cdot(\varepsilon_s\rho_s\overline{u}_s)=0$
Fluid phase momentum equation:
$\frac{\partial}{\partial t}\left(\varepsilon_g\rho_g\overline{u}_g\right)+\nabla\cdot\left(\varepsilon_g\rho_g\overline{u}_g\overline{u}_g\right)=-\varepsilon_g\nabla p_g-\nabla\cdot\left(\varepsilon_g\overline{\overline{\tau}}_g\right)-\beta\left(\overline{u}_g-\overline{u}_s\right)+\varepsilon_g\rho_g\overline{g}$
Solid phase momentum equation:
$\frac{\partial}{\partial t}(\varepsilon_s\rho_s\overline{u}_s)+\nabla\cdot(\varepsilon_s\rho_s\overline{u}_s\overline{u}_s)=-\varepsilon_s\nabla p_g-\nabla p_s-\nabla\cdot(\varepsilon_s\overline{\overline{\tau}}_s)+\beta\left(\overline{u}_g-\overline{u}_s\right)+\varepsilon_s\rho_s\overline{g}$
Granular temperature equation:
$\frac{3}{2}\left[\frac{\partial}{\partial t}(\varepsilon_s\rho_s\theta)+\nabla\cdot(\varepsilon_s\rho_s\theta\overline{u}_s)\right]=-\left(p_s\overline{\overline{I}}+\varepsilon_s\overline{\overline{\tau}}_s\right):\nabla\overline{u}_s-\nabla\cdot(\varepsilon_s\overline{q}_s)-3\beta\theta-\gamma$
Fluid phase thermal energy equation:
$C_{p,g}\left[\frac{\partial}{\partial t}\left(\varepsilon_g\rho_g T_g\right)+\nabla\cdot\left(\varepsilon_g\rho_g\overline{u}_g T_g\right)\right]=\nabla\cdot(\varepsilon_g\kappa_{g,e}\nabla T_g)-\alpha\left(T_g-T_s\right)$
Solid phase thermal energy equation:
$C_{p,s}\left[\frac{\partial}{\partial t}(\varepsilon_s\rho_s T_s)+\nabla\cdot(\varepsilon_s\rho_s\overline{u}_s T_s)\right]=\nabla\cdot(\varepsilon_s\kappa_{s,e}\nabla T_s)+\alpha\left(T_g-T_s\right)$

Table 5 Two-Fluid Model, Closure Equations (Nieuwland et al., 1996)

Fluid–particle drag:

$$\beta = 18\left(1-\varepsilon_g\right)\varepsilon_g\mu_g\frac{F}{d_p^2}$$

Ergun dimensionless drag (1952) for $\varepsilon_g < 0.8$ (Ergun, 1952):

$$F = \frac{150}{18}\frac{\varepsilon_g}{\left(1-\varepsilon_g\right)^2} + \frac{1.75}{18}\frac{Re_p}{\left(1-\varepsilon_g\right)^2} \quad Re_p = \frac{\varepsilon_g\rho_g\left|\overline{u}_g-\overline{u}_s\right|d_p}{\mu_g}$$

Wen and Yu dimensionless drag (1966) for $\varepsilon_g > 0.8$ (Wen and Yu, 1966):

$$F = \frac{1}{24}C_d\varepsilon_g^{-3.65}Re_p$$

$$C_d = \frac{24}{Re_p}\left(1+0.15Re_p^{0.687}\right) \text{ for } Re_p < 1000 \text{ and } C_d = 0.44 \text{ for } Re_p > 1000$$

Fluid–particle heat transfer:

$$\alpha = \frac{6\left(1-\varepsilon_g\right)}{d_p}\alpha_p$$

Gunn (1978) fluid–particle heat transfer correlation:

$$Nu_p = \frac{\alpha_p d_p}{\kappa_g} = \left(7-10\varepsilon_g+5\varepsilon_g^2\right)\left(1+0.7\left(Re_p\right)^{0.2}Pr^{1/3}\right) + \left(1.33-2.4\varepsilon_g+1.2\varepsilon_g^2\right)\left(Re_p\right)^{0.7}Pr^{1/3}$$

$$Pr = \frac{C_{p,g}\mu_g}{\kappa_g}$$

Solid phase pressure:

$$p_s = (1+2(1+e_n)\varepsilon_s g_0)\varepsilon_s\rho_s\theta$$

Newtonian stress tensor fluid phase:

$$\overline{\overline{\tau}}_g = -\left[\left(\lambda_g-\frac{2}{3}\mu_g\right)(\nabla\cdot\overline{u}_g)\overline{\overline{I}} + \mu_g\left((\nabla\overline{u}_g)+(\nabla\overline{u}_g)^T\right)\right]$$

Newtonian stress tensor solid phase:

$$\overline{\overline{\tau}}_s = -\left[\left(\lambda_s-\frac{2}{3}\mu_s\right)(\nabla\cdot\overline{u}_s)\overline{\overline{I}} + \mu_s\left((\nabla\overline{u}_s)+(\nabla\overline{u}_s)^T\right)\right]$$

Bulk viscosity solid phase:

$$\lambda_s = \frac{4}{3}\varepsilon_s\rho_s d_p g_0(1+e_n)\sqrt{\frac{\theta}{\pi}}$$

Shear viscosity solid phase:

$$\mu_s = 1.01600\frac{5}{96}\pi\rho_s d_p\sqrt{\frac{\theta}{\pi}}\frac{\left(1+\frac{8}{5}\frac{(1+e_n)}{2}\varepsilon_s g_0\right)\left(1+\frac{8}{5}\varepsilon_s g_0\right)}{\varepsilon_s g_0} + \frac{4}{5}\varepsilon_s\rho_s d_p g_0(1+e_n)\sqrt{\frac{\theta}{\pi}} + \mu_s^f$$

Pseudo Fourier fluctuating kinetic energy flux:

$$\overline{q}_s = -\kappa_s\nabla\theta$$

Pseudothermal conductivity solid phase:

$$\kappa_s = 1.02513\frac{75}{384}\pi\rho_s d_p\sqrt{\frac{\theta}{\pi}}\frac{\left(1+\frac{12}{5}\frac{(1+e_n)}{2}\varepsilon_s g_0\right)\left(1+\frac{12}{5}\varepsilon_s g_0\right)}{\varepsilon_s g_0} + 2\varepsilon_s\rho_s d_p g_0(1+e_n)\sqrt{\frac{\theta}{\pi}}$$

Continued

Table 5 Two-Fluid Model, Closure Equations (Nieuwland et al., 1996)—cont'd

Dissipation rate of granular energy due to inelastic particle–particle collisions:

$$\gamma = 3\left(1-e_n^2\right)\varepsilon_s^2\rho_s g_0\theta\left[\frac{4}{d_p}\sqrt{\frac{\theta}{\pi}}-\left(\nabla\cdot\vec{u}_s\right)\right]$$

Radial distribution function:

$$g_0 = 1+4\varepsilon_s\frac{1+2.5\varepsilon_s+4.5904\varepsilon_s^2+4.515439\varepsilon_s^2}{\left(1-\left(\frac{\varepsilon_s}{\varepsilon_{s,\max}}\right)^3\right)^{0.67802}}$$

$$\varepsilon_{s,\max} = 0.64356$$

Frictional stress model:

$$\mu_s^f = \frac{p_c\sqrt{2}\sin\phi}{2\varepsilon_s\sqrt{\left(\overline{\overline{D}}:\overline{\overline{D}}+\theta/d_p^2\right)}},\quad \overline{\overline{D}} = \left(\frac{1}{2}\left((\nabla\overline{u}_s)+(\nabla\overline{u}_s)^T\right)-\frac{1}{3}(\nabla\cdot\overline{u}_s)\overline{\overline{I}}\right)$$

$$p_c = \begin{cases} F\dfrac{\left(\varepsilon_s-\varepsilon_{s,\min}\right)^r}{\left(\varepsilon_{s,\max}-\varepsilon_s\right)^s} & \varepsilon_s > \varepsilon_{s,\min} \\ 0 & \varepsilon_s < \varepsilon_{s,\min} \end{cases}$$

$F = 0.05(\mathrm{N/m^2})$, $r = 2$, $s = 3$, $\varepsilon_{s,\min} = 0.5$, $\phi = 28$ degree

Fluid phase thermal conductivity:

$$\kappa_{g,e} = \left(\frac{1-\sqrt{1-\varepsilon_g}}{\varepsilon_g}\right)\kappa_g$$

Solid phase thermal conductivity:

$$\kappa_{s,e} = \left(\frac{\omega A+(1-\omega)\Gamma}{\sqrt{1-\varepsilon_g}}\right)\kappa_s$$

where

$$\Gamma = \frac{2}{\left(1-\frac{B}{A}\right)}\left(\frac{(A-1)}{\left(1-\frac{B}{A}\right)^2}\frac{B}{A}\ln\left(\frac{A}{B}\right)-\frac{(B-1)}{\left(1-\frac{B}{A}\right)}-\frac{1}{2}(B+1)\right)$$

with

$$B = 1.25\left(\frac{1-\varepsilon_g}{\varepsilon_g}\right)^{10/9}\quad\text{and}\quad A = \frac{\kappa_s}{\kappa_g}\quad \omega = 0.00726$$

of granular energy due to inelastic particle–particle collisions is required as well as a closure for the so-called frictional stress model which is important in quasi-static regions of the fluidized bed where the basic assumptions underlying the KTGF break down. The KTGF reported in this chapter is mainly based on Nieuwland et al. (1996) and the extended model (to incorporate frictional stress models) of Verma et al. (2013). Finally for the solution of the complete set of equations suitable boundary conditions need to be specified at all boundaries of the computational domain. For the solid phase, the boundary conditions for confining walls have been formulated by Pita and Sundaresan (1991), respectively, for the solids phase momentum equation and the granular temperature equation:

$$\left(\overline{\overline{I}} - \overline{n}\overline{n}\right)\left(\overline{\overline{\tau}}_s \cdot \overline{n}\right) = \frac{\varphi\sqrt{3}\pi\varepsilon_s\rho_s g_0\theta\overline{u}_s}{6\varepsilon_{s,\max}} \tag{52}$$

$$\left(\overline{q}_s \cdot \overline{n}\right) = \frac{\sqrt{3}\pi\varepsilon_s\rho_s g_0\theta^{3/2}\left(1 - e_w^2\right)}{4\varepsilon_{s,\max}} - \left(\overline{u}_s \cdot \overline{\overline{\tau}}_s \cdot \overline{n}\right) \tag{53}$$

4.2 Numerical Solution Method TFM

For the solution of the governing equations, a semiimplicit methodology is used described in detail in Verma et al. (2013). Consequently only the main features of the method will be given here starting with the time discretization of the continuity and momentum equations:

$$\left(\varepsilon_g\rho_g\right)^{n+1} = \left(\varepsilon_g\rho_g\right)^{n} - \Delta t\left(\nabla \cdot \varepsilon_g\rho_g\overline{u}_g\right)^{n+1} \tag{54}$$

$$\left(\varepsilon_s\rho_s\right)^{n+1} = \left(\varepsilon_s\rho_s\right)^{n} - \Delta t\left(\nabla \cdot \varepsilon_s\rho_s\overline{u}_s\right)^{n+1} \tag{55}$$

$$\begin{aligned}\left(\varepsilon_g\rho_g\overline{u}_g\right)^{n+1} &= \left(\varepsilon_g\rho_g\overline{u}_g\right)^{n} - \Delta t\left(\varepsilon_g\nabla p_g\right)^{n+1} + \overline{A}_g \\ &\quad - \Delta t\beta^n\left(\overline{u}_g - \overline{u}_s\right) + \Delta t\left(\varepsilon_g\rho_g\overline{g}\right)\end{aligned} \tag{56}$$

$$\begin{aligned}\left(\varepsilon_s\rho_s\overline{u}_s\right)^{n+1} &= \left(\varepsilon_s\rho_s\overline{u}_s\right)^{n} - \Delta t\left(\varepsilon_s\nabla p_g\right)^{n+1} \\ &\quad - \Delta t\left(\nabla p_s\right)^{n+1} + \overline{A}_s + \Delta t\beta^n\left(\overline{u}_g - \overline{u}_s\right) + \Delta \mathrm{t}\left(\varepsilon_s\rho_s\overline{g}\right)\end{aligned} \tag{57}$$

where n represents the time index whereas $\overline{A}_g$ and $\overline{A}_s$ contain, respectively, the convective and viscous contributions for the gas phase and solid phase. For the spatial discretization, a uniform staggered computational grid is used where the scalar variables are defined at the cell centers whereas the velocity components are defined at the cell faces.

The viscous contributions are treated with a mixed explicit–implicit scheme where the normal and mixed derivatives are treated, respectively, implicitly and explicitly. In the first step of the solution procedure intermediate momentum densities are computed for both phases on basis of the following two equations where the pressure and interphase momentum exchange terms in Eqs. (56) and (57) are excluded:

$$\left(\varepsilon_g\rho_g\overline{u}_g\right)^{**} = \left(\varepsilon_g\rho_g\overline{u}_g\right)^{n} + \overline{A}^{**}_{g,\mathrm{con+vis}} + \overline{A}^{n}_{g,\mathrm{con+vis}} + \Delta \mathrm{t}\left(\varepsilon_g\rho_g\overline{g}\right) \tag{58}$$

$$\left(\varepsilon_s\rho_s\overline{u}_s\right)^{**} = \left(\varepsilon_s\rho_s\overline{u}_s\right)^{n} + \overline{A}^{**}_{s,\mathrm{con+vis}} + \overline{A}^{n}_{s,\mathrm{con+vis}} + \Delta \mathrm{t}\left(\varepsilon_s\rho_s\overline{g}\right) \tag{59}$$

The convective contributions are treated with the deferred correction approach where the convective fluxes are written as follows:

$$\begin{aligned} A^{**}_{\mathrm{g,con}} &= \overline{A}^{n+1}_{\mathrm{g,con,UPW}} + \left(\overline{A}^{n}_{\mathrm{g,con,TVD}} - \overline{A}^{n}_{\mathrm{g,con,UPW}}\right) \\ A^{**}_{\mathrm{s,con}} &= \overline{A}^{n+1}_{\mathrm{s,con,UPW}} + \left(\overline{A}^{n}_{\mathrm{s,con,TVD}} - \overline{A}^{n}_{\mathrm{s,con,UPW}}\right) \end{aligned} \tag{60}$$

where the lower-order upwind (UPW) part is treated implicitly whereas the higher-order part (based on total variation diminishing (TVD)) is treated explicitly. This approach contributes to the main diagonal dominance of the algebraic equations and provides a stable solution with better convergence characteristics. Subsequently these intermediate momentum densities are corrected by solving the following two equations:

$$\left(\varepsilon_{\mathrm{g}}\rho_{\mathrm{g}}\overline{u}_{\mathrm{g}}\right)^{n+1} = \left(\varepsilon_{\mathrm{g}}\rho_{\mathrm{g}}\overline{u}_{\mathrm{g}}\right)^{**} - \Delta t\left(\varepsilon_{\mathrm{g}}\nabla p_{\mathrm{g}}\right)^{n+1} - \Delta t\beta^{n}\left(\overline{u}_{\mathrm{g}} - \overline{u}_{\mathrm{s}}\right) \tag{61}$$

$$\begin{aligned} \left(\varepsilon_{\mathrm{s}}\rho_{\mathrm{s}}\overline{u}_{\mathrm{s}}\right)^{n+1} = {} & \left(\varepsilon_{\mathrm{s}}\rho_{\mathrm{s}}\overline{u}_{\mathrm{s}}\right)^{**} - \Delta t\left(\varepsilon_{\mathrm{s}}\nabla p_{\mathrm{g}}\right)^{n+1} \\ & - \Delta t\left(\nabla p_{\mathrm{s}}\right)^{n+1} + \Delta t\beta^{n}\left(\overline{u}_{\mathrm{g}} - \overline{u}_{\mathrm{s}}\right) \end{aligned} \tag{62}$$

where the advanced time level pressure and solids volume fraction are computed from the following two equations for the respective corrections:

$$\begin{aligned} & \left(J_{\mathrm{g}}\right)^{n}_{i-1,j,k}\left(\Delta p_{\mathrm{g}}\right)^{n+1}_{i-1,j,k} + \left(J_{\mathrm{g}}\right)^{n}_{i+1,j,k}\left(\Delta p_{\mathrm{g}}\right)^{n+1}_{i+1,j,k} + \left(J_{\mathrm{g}}\right)^{n}_{i,j-1,k}\left(\Delta p_{\mathrm{g}}\right)^{n+1}_{i,j-1,k} \\ & + \left(J_{\mathrm{g}}\right)^{n}_{i,j+1,k}\left(\Delta p_{\mathrm{g}}\right)^{n+1}_{i,j+1,k} + \left(J_{\mathrm{g}}\right)^{n}_{i,j,k-1}\left(\Delta p_{\mathrm{g}}\right)^{n+1}_{i,j,k-1} \\ & + \left(J_{\mathrm{g}}\right)^{n}_{i,j,k+1}\left(\Delta p_{\mathrm{g}}\right)^{n+1}_{i,j,k+1} + \left(J_{\mathrm{g}}\right)^{n}_{i,j,k}\left(\Delta p_{\mathrm{g}}\right)^{n+1}_{i,j,\mathrm{k}} = -\left(D_{\mathrm{g}}\right)^{n+1}_{j,j,k} \end{aligned} \tag{63}$$

$$\begin{aligned} & \left(J_{\mathrm{s}}\right)^{n}_{i-1,j,k}\left(\Delta \varepsilon_{\mathrm{s}}\right)^{n+1}_{i-1,j,k} + \left(J_{\mathrm{s}}\right)^{n}_{i+1,j,k}\left(\Delta \varepsilon_{\mathrm{s}}\right)^{n+1}_{i+1,j,k} + \left(J_{\mathrm{s}}\right)^{n}_{i,j-1,k}\left(\Delta \varepsilon_{\mathrm{s}}\right)^{n+1}_{i,j-1,k} \\ & + \left(J_{\mathrm{s}}\right)^{n}_{i,j+1,k}\left(\Delta \varepsilon_{\mathrm{s}}\right)^{n+1}_{i,j+1,k} + \left(J_{\mathrm{s}}\right)^{n}_{i,j,k-1}\left(\Delta \varepsilon_{\mathrm{s}}\right)^{n+1}_{i,j,k-1} + \left(J_{\mathrm{s}}\right)^{n}_{i,j,k+1}\left(\Delta \varepsilon_{\mathrm{s}}\right)^{n+1}_{i,j,k+1} \\ & + \left(J_{\mathrm{s}}\right)^{n}_{i,j,k}\left(\Delta \varepsilon_{\mathrm{s}}\right)^{n+1}_{i,j,\mathrm{k}} = -\left(D_{\mathrm{s}}\right)^{n+1}_{j,j,k} \end{aligned} \tag{64}$$

where J_{g} and J_{s} represent the Jacobi matrices for, respectively, the gas and solid phase which are evaluated at the old time level. These matrices contain the derivatives of the mass residuals D_{g} and D_{s} according to the continuity equations of the gas and solid phase with respect to the pressure and solids volume fraction, respectively, and can be obtained analytically from the continuity and momentum equations. Effectively the solution method constitutes a two-step projection method combined with an extension of the SIMPLE algorithm of Patankar and Spalding (1972).

Once the corrections for the pressure and solids volume fraction have been obtained from Eqs. (63) and (64), Eqs. (61) and (62) can be solved to obtain new estimates of the advanced time level momentum densities and updated values of the mass residuals D_g and D_s. This process is continued until the mass residuals satisfy the imposed convergence criteria. Following the solution of the continuity and momentum equations, the granular temperature equation is solved by an iterative procedure. Following the solution of the hydrodynamic equations, the thermal energy equations are solved using a fully implicit scheme. The resulting linear set of equations for the intermediate momentum densities (Eqs. 58 and 59), the pressure and solids volume fraction corrections (Eqs. 63 and 64), the discretized granular temperature equation, and the coupled thermal energy equations (see Table 4) are all solved with a sparse matrix solver based on the ICGM.

4.3 Results TFM

In this section, illustrative computational results will be presented highlighting the capabilities of continuum modelling of dense gas-fluidized beds with focus on the hydrodynamics (gas bubbles behavior, solids motion, and mixing and heat transfer). In Section 4.3.1, we will focus on pseudo 2D beds whereas in Section 4.3.2, we will focus on 3D cylindrical beds. Finally in Section 4.3.3, we will focus on nonisothermal systems involving a 3D cylindrical gas-fluidized with heat production in the solid phase.

4.3.1 Hydrodynamics of a Pseudo 2D Gas-Fluidized Bed

Pseudo 2D gas-fluidized beds have been used extensively to study in detail the hydrodynamics because the complex behavior of the rising gas bubbles and associated solids motion can be monitored with relative ease using powerful noninvasive optical techniques such as DIA and PIV. In Fig. 12, we show the bubble formation and propagation computed from the TFM governing equations for a pseudo 2D gas-fluidized bed with a central jet. The data used for this simulation are given in Table 6.

The simulation results reveal a complex pattern of bubble formation with rapid growth of the bubbles due to strong vertical coalescence of the detached bubbles. A similar pattern is observed in experiments and it has been shown that the initial bubble size and time-averaged porosity distribution obtained from similar computations agree reasonably well with experimental data obtained for such pseudo 2D gas-fluidized beds (see, respectively, Kuipers et al., 1991, 1992a). These studies also revealed that during the initial stage of bubble formation at an orifice substantial leakage

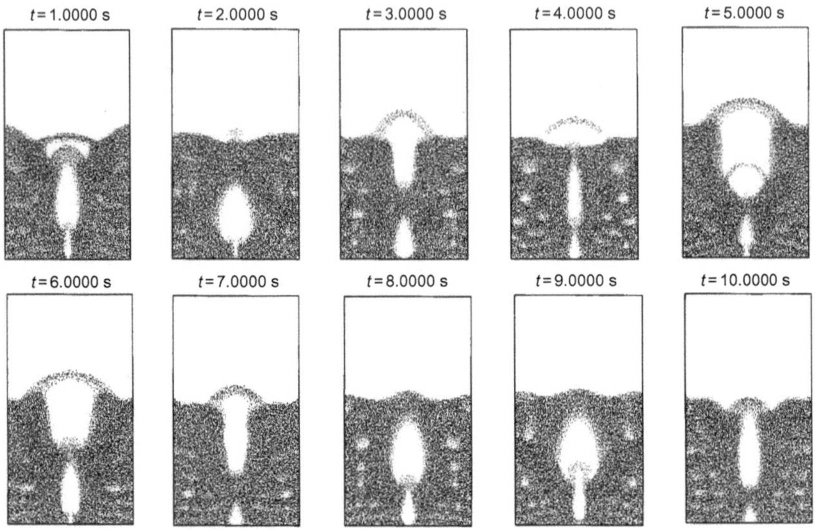

Fig. 12 Bubble formation and propagation in a 2D gas-fluidized bed with a central jet. See Table 3 for the data used in the simulation. Visualization of the porosity distribution is done with "dot plots" where the number of dots plotted for a specific computational cell is proportional to the solids volume fraction. For cells with a solids volume fraction less than 0.15 no dots are distributed.

Table 6 Data Used for the Numerical Simulation of Bubble Formation and Propagation in a Pseudo 2D Gas-Fluidized Bed With a Central Jet

Gas density	Ideal gas law	kg/m^3
Gas viscosity	2.0×10^{-5}	kg/(m s)
Particle size	500	μm
Particle density	2660	kg/m^3
Restitution coefficient	0.95	(–)
Time step	10^{-4}	s
Bed dimensions	0.570 (x-direction) 0.015 (y-direction) 1.000 (z-direction)	m m m
Computational grid size and number of computational cells	0.0075 (x-direction) (76 cells) 0.0075 (y-direction) (2 cells) 0.0100 (z-direction) (100 cells)	m m m
Background velocity	0.25	m/s
Jet velocity	10.0	m/s

of the gas injected through the orifice into the emulsion phase occurs which is particularly important in case fast heterogeneously catalyzed reactions take place.

With the advent of powerful optical measuring techniques such as DIA and PIV the solids motion in such systems could be studied in detail as well and is shown in Fig. 13. In this figure, the time-averaged filtered emulsion phase velocity measured in a bed with a width of 0.30 m and a packed height of 0.30 m is shown for several fluidization velocities.

The combined usage of DIA and PIV allows to correct for the influence of particle raining through the roof of the bubbles on the time-averaged emulsion phase velocities (Laverman et al., 2008). From this figure, it can be seen that the solids motion is quite complex with circulation cells in the top part of the bed where the strength of the circulation cells depends strongly on the fluidization velocity. In the center of the bed the particles flow upward due to the strong bubble propagation in this part of the bed, while the particles flow downward near the confining walls. These datasets have been used extensively to validate both DEM and TFM models for gas-fluidized beds.

4.3.2 Hydrodynamics of a 3D Cylindrical Bed

In the previous section, the focus was on pseudo 2D gas-fluidized beds which feature many important properties of real (complex) 3D gas-fluidized beds. However, gas-fluidized beds encountered in the process industries typically possess a cylindrical geometry which introduces an additional degree of freedom for bubble propagation and solids motion. Moreover, internals immersed in the fluidized solids might be present to facilitate efficient heat removal from the bed in case strongly exothermic catalyzed reactions occur. Finally multiple inlets for the gas and/or solids phase might exist to enable continuous and flexible operation (see Fig. 1).

The treatment of the boundary condition at the center of the bed requires a special procedure detailed in Verma et al. (2013) to avoid artificial accumulation of solids near the center line of the bed. As an alternative to solving the TFM conservation equations in full 3D cylindrical coordinates these equations can be solved as well on curvilinear coordinates (Manger, 1996) or on regular Cartesian grids where the cylindrical wall is treated with the IBM. See the excellent review paper by Mittal and Iaccarino (2005) for a more detailed discussion of the latter technique.

Verma et al. (2013) also studied the effect of the discretization scheme for the convection terms and recommended to use high-order schemes such as

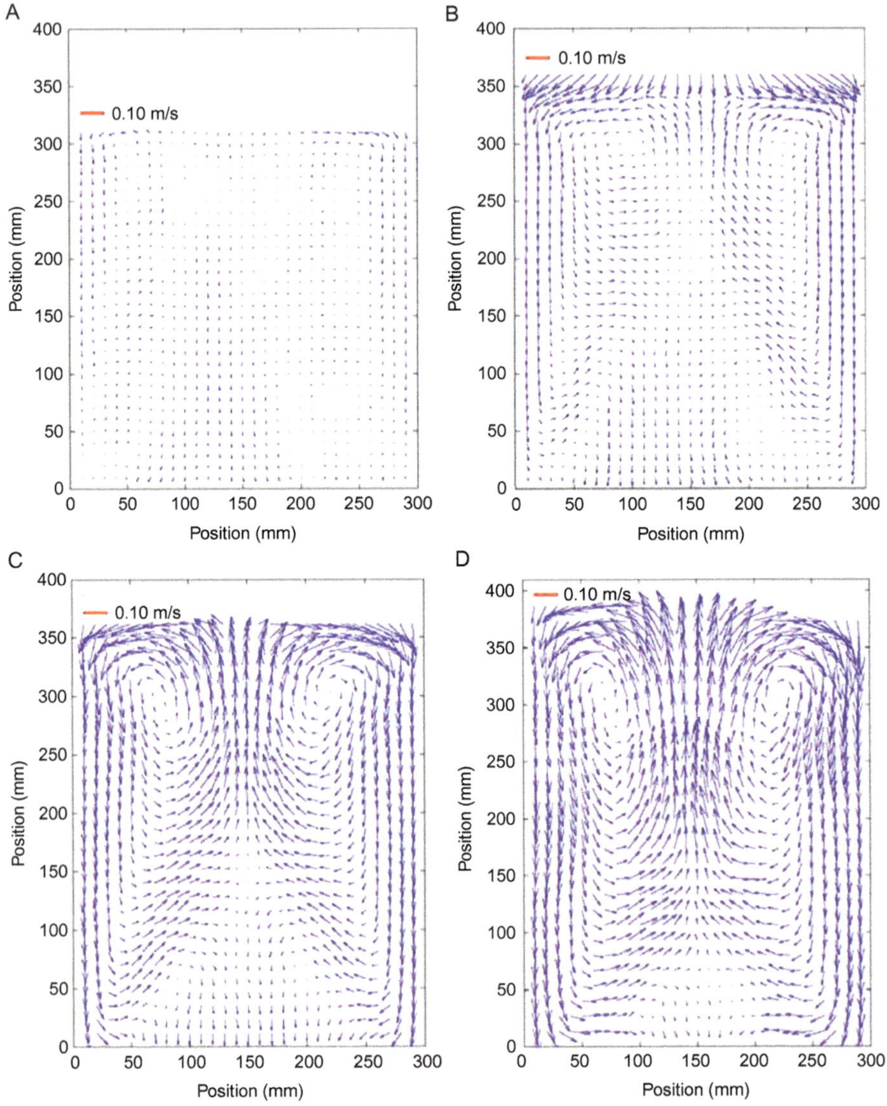

Fig. 13 Time-averaged filtered emulsion phase velocity profiles measured in a bed with a width of 0.30 m and a packed bed ratio of 1.0 for several fluidization velocities: (A) $(u_0/u_{mf}) = 1.5$, (B) $(u_0/u_{mf}) = 2.0$, (C) $(u_0/u_{mf}) = 2.5$, and (D) $(u_0/u_{mf}) = 3.5$. Glass bead in the size range of 400–600 μm (Geldart B-type) were used in these experiments using air as fluidizing agent. *Reproduced with permission from Laverman JA, Roghair I, van Sint Annaland M, Kuipers JAM: Investigation into the hydrodynamics of gas–solid fluidized beds using particle image velocimetry coupled with digital image analysis,* Can J Chem Eng *86:523–535, 2008.*

TVD schemes instead of the first-order upwind (FOU) scheme to avoid excessive numerical diffusion or "computational smearing." This is especially important to capture the small bubbles formed in the bottom zone of the bed. Moreover, for dense gas-fluidized beds with heterogeneously catalyzed reactions the preservation of sharp boundaries between gas bubbles and the dense emulsion phase is of crucial importance to correctly predict the degree of chemical conversion in case fast reactions proceed. The data used for the numerical simulation of a freely bubbling 3D cylindrical bed are summarized in Table 7. In Fig. 14, we show instantaneous snapshots of the volume fraction of the gas in the central plane of a 3D 0.306 m diameter gas-fluidized bed revealing a dynamic evolution of a very complex void structure.

From Fig. 14, it can be seen that the small bubbles that form near the gas distributor plate rapidly coalesce to form larger bubbles at a higher elevation in the bed. From animations of this process, it can clearly be seen that these large bubbles tend to rise in the center of the bed. This pattern of bubble propagation is well known and leads to a large-scale circulation pattern in the bed with solids up-flow in the central portion of the bed and solids down-flow near the cylindrical wall of the bed.

Table 7 Data Used for the Numerical Simulation of a Freely Bubbling 3D Cylindrical Dense Gas-Fluidized Bed

Gas density	Ideal gas law	kg/m^3
Gas viscosity	2.0×10^{-5}	kg/(m s)
Particle size	500	μm
Particle density	2500	kg/m^3
Restitution coefficient	0.86	(–)
Time step	10^{-4}	s
Bed dimensions	0.306 (radial direction) 1.000 (axial direction)	m m
Computational grid	45 (radial direction) 30 (azimuthal direction) 240 (axial direction)	(–) (–) (–)
Minimum fluidization velocity U_{mf}	0.18	m/s
Superficial gas velocity	$3.5U_{mf}$	m/s

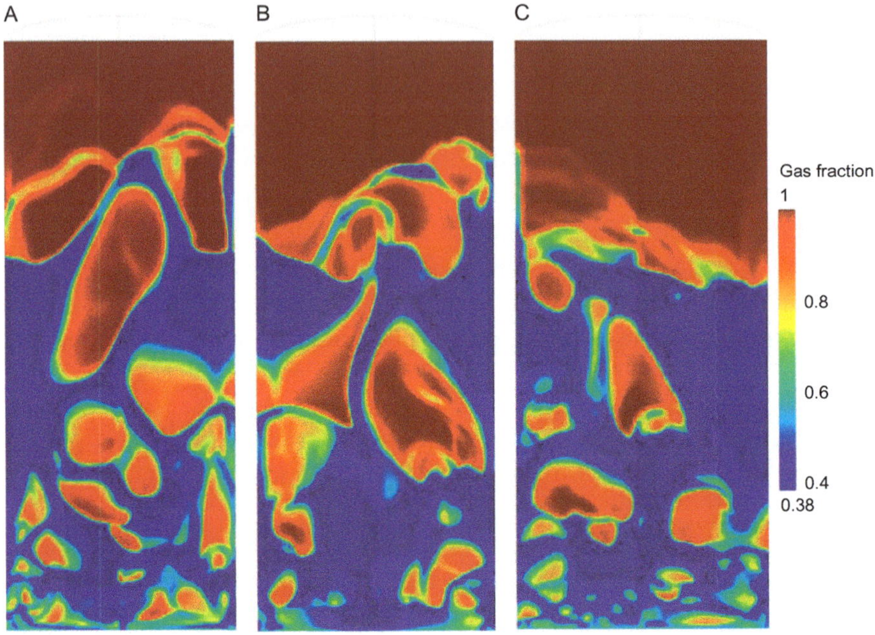

Fig. 14 Instantaneous snapshots of the volume fraction of gas in the central plane of a 3D 0.306 m diameter gas-fluidized bed at $t = 2.6$ s (A), $t = 4.0$ s (B), and $t = 5.0$ s (C). The simulation settings are given in Table 7. *Reproduced with permission from Verma V, Deen NG, Padding JT, Kuipers JAM: Two-fluid modelling of three-dimensional cylindrical gas–solid fluidized beds using the kinetic theory of granular flow,* Chem Eng Sci *102:227–245, 2013.*

This pattern of solids motion is also found experimentally and as an example we shown in Fig. 15, the azimuthally and time-averaged solids circulation pattern obtained from experiments (Laverman et al., 2012) and TFM simulations. The experimental results were obtained with a positron emission particle tracking (PEPT) technique in which the position of an activated particle was measured as a function of time over a sufficiently long measuring time. Inspection of Fig. 15 reveals the presence of two circulation cells, one in the bottom portion of the bed and a larger cell in the upper portion of the bed. The main features of the experimentally found solids circulation pattern are very well reproduced by the TFM simulations. Verma et al. (2015) also studied the effect of bed size and aspect ratio on the hydrodynamics in 3D gas–solid fluidized beds with Geldart B-type particles and found that the bubble size closely follows the empirical correlation proposed by Werther (1976) for small beds and the correlation proposed by Darton et al. (1977) for sufficiently large beds. They also found that the bubble size

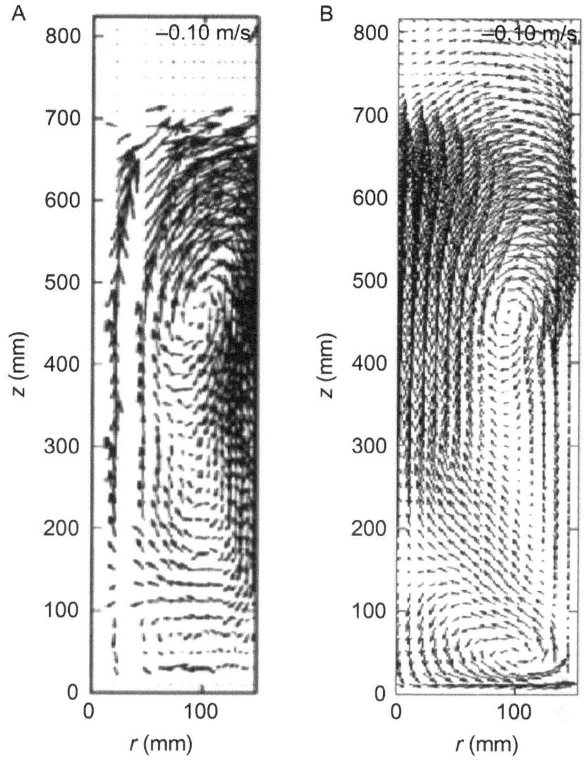

Fig. 15 Azimuthally and time-averaged solids circulation pattern obtained from experiments using PEPT (Laverman et al., 2012) and TFM simulations. See Table 7 for the data used in the simulation. *Reproduced with permission from Verma V, Deen NG, Padding JT, Kuipers JAM: Two-fluid modelling of three-dimensional cylindrical gas–solid fluidized beds using the kinetic theory of granular flow,* Chem Eng Sci *102:227–245, 2013.*

increases when the bed diameter is increased from 0.10 to 0.30 m diameter and remains approximately constant for bed diameters ranging from 0.30 to 1.0 m. For the largest bed diameter of 1.0 m, they found an increase in the bubble rise velocity which was attributed to the so-called Gulf Stream circulation which becomes important for large diameter beds (Matsen, 1996). This phenomenon leads to a concentration of the gas bubbles in the center of the bed and significantly affects the rise velocity of the bubbles.

The empirical correlations of Werther (1976) and Darton et al. (1977) are based on point measurements using so-called probe techniques. Such measurements are very time consuming and as such TFM simulations provide an attractive alternative to obtain data on bubble size. Verma et al.

(2015) also found a pronounced difference in shallow and deep beds with respect to the bubble size and solids circulation patterns which is expected behavior in view of the opportunity of the rising gas bubbles to coalesce.

Besides the comparison of the simulated bubble size with data obtained from well-established correlations, Verma et al. (2014b) also compared their simulation results on bubble dynamics in a 3D gas-fluidized bed with experimental data obtained using ultrafast electron beam X-ray tomography and reported good agreement between the bubble sizes obtained from experiments and TFM simulations as shown in Fig. 16.

Verma et al. (2014a) also studied the effect of operating pressure on the hydrodynamics of a 0.30 m diameter gas-fluidized bed containing 1.1 mm diameter LLDPE particles using a constant excess gas velocity. They found a substantial effect of pressure on bubble size and shape (up to 8 bar for shape), pressure fluctuations, the time-averaged porosity distribution, and the solids circulation pattern. They reported that the bubble size decreases with increasing operating pressure and that the distinct two-phase flow structure becomes less pronounced with increasing pressure. In other words, these results indicate that gradually the heterogeneous flow structure disappears with increasing operating pressure. In Fig. 17, we show the PDF of the time-averaged porosity distribution for several heights in the bed at atmospheric pressure (left) and an elevated pressure of 20 bar (right). It can clearly be seen that the PDF becomes flatter with increasing pressure where the prevalence of intermediate porosity values (i.e., ranging from 0.6 to 0.8) is clearly noticeable.

4.3.3 Gas-Fluidized Bed With Heat Production

In many processes involving gas-fluidized beds the solid phase acts as a heterogeneous catalyst where significant heat effects play a role. Combustion, gasification, and gas phase polymerization for the production of polyolefines can be mentioned as specific examples. In the latter process, the monomer (ethylene or propylene) is contacted with a prepolymerized particles containing metallocene type of catalysts leading to rapid growth of the particles with liberation of a large amount of heat due to the strongly exothermic polymerization reactions. In this type of application, it is particularly important to avoid excessive local heating of the polymer particles (hot spot formation) because it could lead to the formation of large agglomerates causing massive defluidization and necessary shutdown of the process. The experimental investigation of such systems is very complicated and costly and as such (validated) hydrodynamic models of gas-fluidized beds

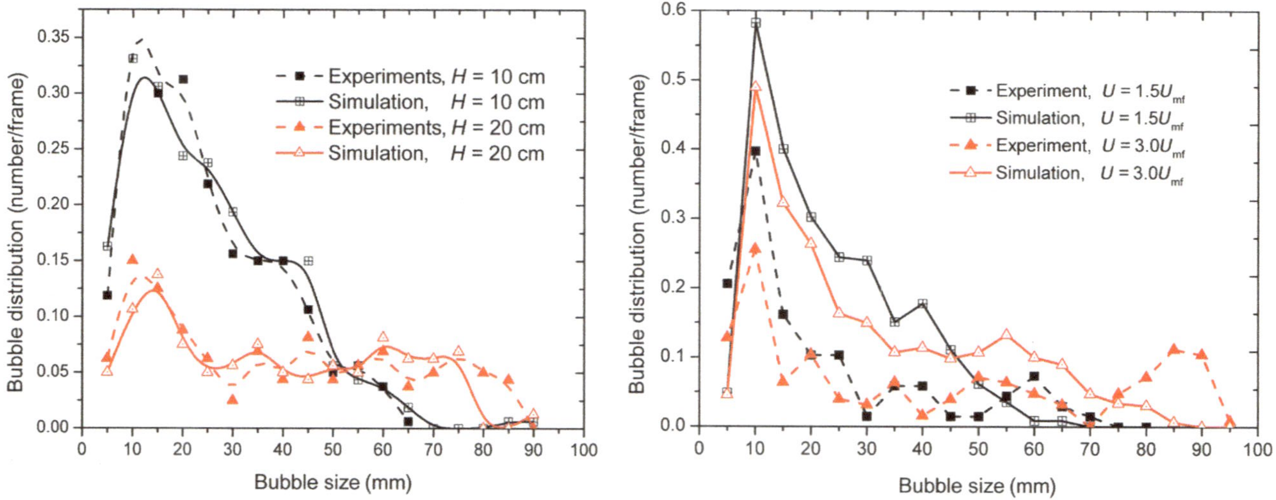

Fig. 16 Comparison of bubble size distributions computed by TFM and measured by X-ray tomography. (*Left*) LLDPE particles at an inlet gas velocity of 1.5 U_{mf} two height. (*Right*) Alumina particles at a height of 10 cm for two gas inlet velocities. *Reproduced with permission from Verma V, Padding JT, Deen NG, Kuipers JAM, Barthel F, Bieberle M, Wagner M, Hampel U: Bubble dynamics in a 3D gas–solid fluidized bed using ultrafast electron beam X-ray tomography and two-fluid model,* AIChE J *60(5):1632–1644, 2014b.*

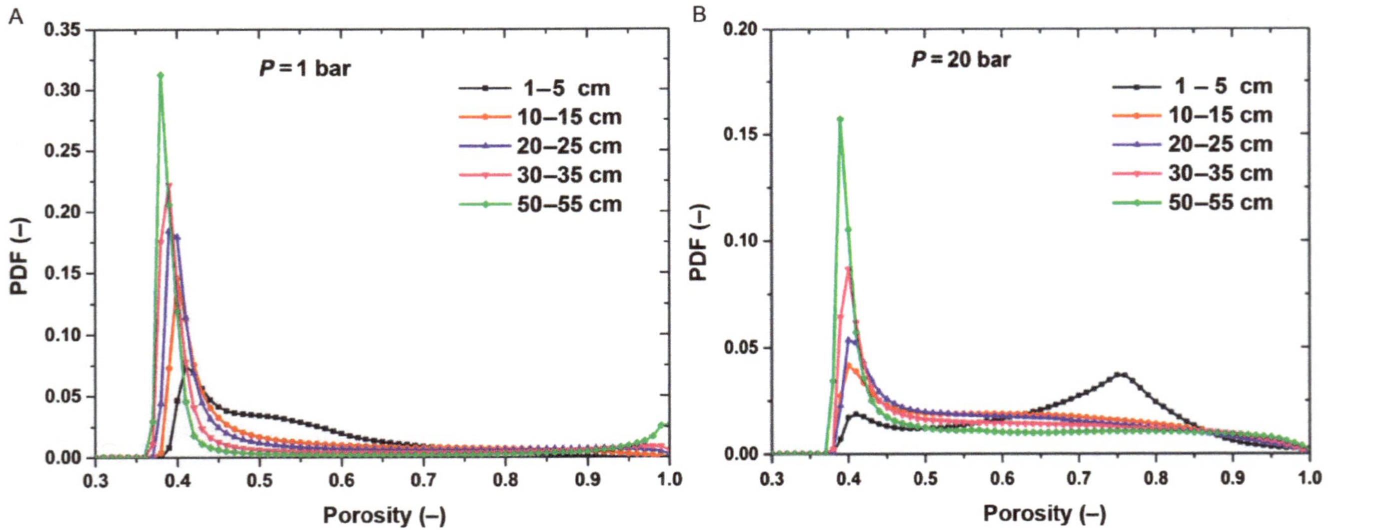

Fig. 17 PDF of the time-averaged porosity distribution for several heights in the bed at atmospheric pressure (A) and an elevated pressure of 20 bar (B). Bed and particle diameter: 0.30 m and 1.1 mm. *Reproduced with permission from Verma V, Padding JT, Deen NG, Kuipers JAM: Numerical investigation on the effect of pressure on fluidization in a 3D fluidized bed,* I&EC Res *53:17487–17498, 2014a.*

supplemented with thermal energy equations offer an attractive alternative to study the complex thermal behavior.

Banaei et al. (2017) modified the thermal energy equation for the solid phase to account for the heat production in the particles by including a constant volumetric heat source (see Table 4). The value of this source was estimated from conditions which typically are encountered in gas phase polymerization processes (see Table 8). Following a careful verification of their model using both analytical solutions for unsteady 1D heat conduction with heat production and results from independent numerical solutions obtained for transient heat transfer in a fixed bed with heat production, they studied the thermal behavior of gas-fluidized beds with heat production.

Table 8 Data Used for the Numerical Simulation of a Gas-Fluidized Bed With Heat Production in the Solid Phase

Gas density	Ideal gas law	kg/m^3
Molar weight of gas	42.08	kg/kmol
Gas viscosity	1.0×10^{-5}	kg/(m s)
Gas thermal conductivity	0.0209	W/(m K)
Gas heat capacity	1670.0	J/(kg K)
Gas inlet temperature	324.0	K
Particle size	0.9875	mm
Particle density	667.0	kg/m^3
Particle–particle restitution coefficient	0.60	(–)
Particle–wall restitution coefficient	0.93	(–)
Specularity coefficient	0.50	(–)
Solid thermal conductivity	1.04	W/(m K)
Solid heat capacity	1670.0	J/(kg K)
Solid volumetric heat production rate	6.7×10^5	W/m^3
Time step	2.0×10^{-6}	s
Bed dimensions	0.06 (radial direction) 0.24 (axial direction)	m m
Computational grid	6 (radial direction) 32 (azimuthal direction) 48 (axial direction)	(–) (–) (–)

Building on the hydrodynamic model developed and validated by Verma et al. (2013, 2014a, 2014b, 2015) they studied the effect of superficial gas velocity on the solid temperature distribution in gas-fluidized beds with heat production in the context of gas phase polymerization of olefins. The data used for the numerical simulations are given in Table 8. The initial conditions for the thermal energy equations were determined by assuming that the gas-fluidized bed behaves like a continuously stirred tank reactor (CSTR). This approach avoids performing very time-consuming TFM simulations that would otherwise be required to reach the thermal steady state of the system.

In Fig. 18, we show the effect of superficial gas velocity on the solid temperature distribution in the bed revealing a rather narrow distribution where the average and spread of the distribution decreases with increasing superficial velocity. These results indicate that the temperature of the solid phase becomes more uniform with increasing superficial gas velocity due to the increased bubble-induced agitation of the solid phase.

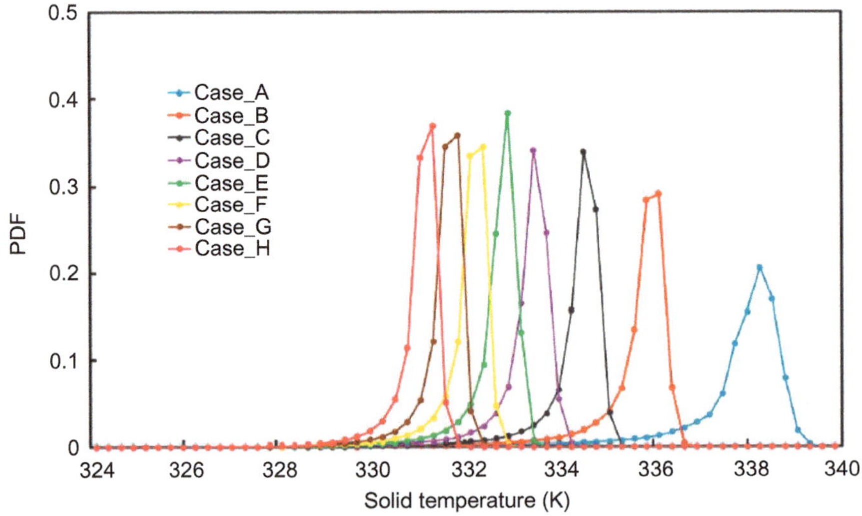

Fig. 18 Effect of superficial gas velocity on the solid temperature distribution. Case_A: $U_0 = 0.4$ m/s, Case_B: $U_0 = 0.5$ m/s, Case_C: $U_0 = 0.6$ m/s, Case_D: $U_0 = 0.7$ m/s, Case_E: $U_0 = 0.8$ m/s, Case_F: $U_0 = 0.9$ m/s, Case_G: $U_0 = 1.0$ m/s, and Case_H: $U_0 = 1.1$ m/s. See Table 8 for the data used in the simulations. *Reproduced with permission from Banaei M, Jegers J, van Sint Annaland M, Kuipers JAM, Deen NG: Effect of superficial gas velocity on the solid temperature in gas fluidized beds with heat production,* I&EC Res *56:8729–8737, 2017.*

5. CONCLUSIONS AND OUTLOOK

In this chapter, the multiscale modelling approach of dense gas–particle flows has been discussed. These flows are extremely complex due to the fact that phenomena need to be considered which prevail over a wide range of time and length scales.

Dense gas–particle flows are frequently encountered in large-scale industrial processes involving granulation, coating, and production of base chemicals and polymers. Both (effective) fluid–particle and (dissipative) particle–particle interaction need to be accounted for because these phenomena and their mutual competition govern the key features of dense gas-fluidized beds such as the occurrence of (large-scale) heterogeneous flow structures which have a profound impact on the performance of dense gas-fluidized bed reactors.

In this chapter, it has been shown that DNS offers a viable alternative in comparison to extensive and time-consuming experimentation to obtain closures for the exchange of mass, momentum, and heat in dense gas–particle flows. The availability of such closures is of vital importance for the accurate simulation of dense particle laden flows using DEM and full continuum models. Despite the fact that considerable progress has been made in recent years enormous challenges remain to expand the multiscale approach to systems involving more complex physics. Systems involving poly-disperse dense gas–particle flows have received very limited attention so far. First of all, accurate closures for the exchange of mass, momentum, and heat are not available for dense gas–particle flows and here DNS offers a lot of potential. In addition, further development of CFD–DEM models for nonspherical particles (including particulates with sharp edges) is highly desired.

Modern continuum models of gas-fluidized beds are based on the KTGF which can be considered as a generalization of the theoretical framework underlying the "classical" continuum model for dense gas–particle flow originally proposed by Anderson and Jackson (1967). The governing equations have been presented together with the closure equations and the main features of the numerical solution method. For KTGF-based continuum models, further developments are necessary to include tangential friction as well as the development of efficient computational schemes to solve the strongly coupled highly nonlinear equations. Recently Yang et al. (2016a, 2016b) extended the "classical" TFM based on the KTGF to include tangential friction for particle–particle collisions as well as particle–wall collisions (Yang et al., 2017b) and validated their model using magnetic particle tracking experiments and discrete particle simulations (Yang et al., 2017a). Moreover the incorporation of tangential friction in multifluid models (Manger, 1996; Van Sint Annaland et al., 2009a, 2009b) is needed.

In certain applications, such as coating, granulation, and gas phase polymerization so-called wet collisions occur which are strongly dissipative. The extension of the KTGF-based continuum models to such systems poses a big challenge since the KTGF framework relies on the prevalence of "slightly inelastic collisions." Finally extensions are required to include simultaneous intraparticle transport of species and heat accompanied by physical and/or chemical transformations.

ACKNOWLEDGMENTS

The authors would like to acknowledge the European Research Council (ERC) for its generous financial support, under its Advanced Investigator Grant scheme, contract number 247298 (MultiScaleFlows).

REFERENCES

Anderson TB, Jackson R: A fluid mechanical description of fluidized beds, *Ind Eng Chem Fundamen* 6:527–534, 1967.

Banaei, M., Jegers, J., van Sint Annaland, M., Kuipers, J.A.M. and Deen, N.G., 2017, Effect of superficial gas velocity on the solid temperature in gas fluidized beds with heat production Ind Eng Chem Res, 56, pp 8729–8737.

Bird RB, Stewart WE, Lightfoot EN: *Transport phenomena,* UK, 1960, John Wiley & Sons.

Buck B, Tang Y, Heinrich S, Deen NG, Kuipers JAM: Collision dynamics of wet solids: rebound and rotation, *Powder Technol* 316:218–224, 2017.

Carlos Varas AE, Peters EAJF, Deen NG, Kuipers JAM: Solids volume fraction measurements on riser flow using a temporal-histogram based DIA method, *AIChE J* 62(8):2681–2698, 2016.

Carlos Varas AE, Peters EAJF, Kuipers JAM: CFD–DEM simulations and experimental validation of clustering phenomena and riser hydrodynamics, *Chem Eng Sci* 169:246–258, 2017a.

Carlos Varas AE, Peters EAJF, Kuipers JAM: Computational fluid dynamics—discrete element method (CFD–DEM) study of mass transfer mechanisms in riser flow, *Ind Eng Chem Res* 56:5558–5572, 2017b.

Centrella J, Wilson J: Planar numerical cosmology. II. The difference equations and numerical tests, *Astrophysical J Suppl Ser* 54:229, 1984.

Darton RC, La Naueza RD, Davidson JF, Harrison D: Bubble growth due to coalescence in fluidized beds, *Trans Inst Chem Eng* 55:274–280, 1977.

Das S, Deen NG, Kuipers JAM: Direct numerical simulation for flow and heat transfer through random open-cell solid foams: development of an IBM based CFD model, *Catal Today* 273:140–150, 2016.

Das S, Deen NG, Kuipers JAM: A DNS study of flow and heat transfer through slender fixed-bed reactors randomly packed with particles, *Chem Eng Sci* 160:1–19, 2017.

Davidson JF: Symposium on fluidization—discussion, *Trans Inst Chem Eng* 39:230–232, 1961.
Deen NG, Kuipers JAM: Direct numerical simulation of fluid flow and mass transfer in dense fluid–particle systems, *Ind Eng Chem Res* 52:11266–11274, 2013.
Deen NG, Kuipers JAM: Direct numerical simulation of fluid flow accompanied by coupled mass and heat transfer in dense fluid–particle systems, *Chem Eng Sci* 116:645–656, 2014.
Deen NG, van Sint Annaland M, van der Hoef MA, Kuipers JAM: Review of discrete particle modelling of fluidized beds, *Chem Eng Sci* 62:28–44, 2007.
Deen NG, Kriebitzsch SHL, van der Hoef MA, Kuipers JAM: Direct numerical simulation of flow and heat transfer in dense fluid–particle systems, *Chem Eng Sci* 81:329–344, 2012.
Deen NG, Peters EAJF, Padding JT, Kuipers JAM: Review of direct numerical simulation of fluid–particle mass, *Chem Eng Sci* 116:710–724, 2014.
Ding J, Gidaspow D: A bubbling fluidization model using kinetic theory of granular flow, *AIChE J* 36(4):523–538, 1990.
Ergun S: Fluid flow through packed columns, *Chem Eng Progress* 48:89–94, 1952.
Gidaspow D: Hydrodynamics of fluidization and heat transfer, *Appl Mech Rev* 39(1):1–23, 1986.
Gunn DJ: Transfer of mass and heat to particles in fixed and fluidized beds, *Int J Heat Mass Transf* 21:467–476, 1978.
Hoomans BPB: *Granular dynamics of gas–solid two-phase flows,* PhD Thesis, The Netherlands, 2000, Twente University.
Hoomans BPB, Kuipers JAM, Briels WJ, Van Swaaij WPM: Discrete particle simulation of bubble and slug formation in a two-dimensional gas-fluidised bed: a hard-sphere approach, *Chem Eng Sci* 51(1):99–118, 1996.
Kuipers JAM, Prins W, van Swaaij WPM: Theoretical and experimental bubble formation at a single orifice in a two-dimensional gas-fluidized bed, *Chem Eng Sci* (11):2881–2894, 1991.
Kuipers JAM, Tammes H, Prins W, van Swaaij WPM: Experimental and theoretical porosity profiles in a two-dimensional gas-fluidized bed with a central jet, *Powder Technol* 71:87–99, 1992a.
Kuipers JAM, van Duijn KJ, van Beckum FPH, van Swaaij WPM: A numerical model of gas-fluidized beds, *Chem Eng Sci* (8):1913–1924, 1992b.
Kunii D, Levenspiel O: *Fluidization engineering,* ed 2, 1991, Butterworth-Heinemann, ISBN: 9780409902334.
Laverman JA, Roghair I, van Sint Annaland M, Kuipers JAM: Investigation into the hydrodynamics of gas–solid fluidized beds using particle image velocimetry coupled with digital image analysis, *Can J Chem Eng* 86:523–535, 2008.
Laverman JA, Fan X, Ingram A, et al: Experimental study on the influence of bed material on the scaling of solids circulation patterns in 3D bubbling gas–solid fluidized beds of glass and polyethylene using positron emission particle tracking, *Powder Technol* 224:297–305, 2012.
Li Z, Janssen TCE, Buist KA, Deen NG, van Sint Annaland M, Kuipers JAM: Experimental and simulation study of heat transfer in fluidized beds with heat production, *Chem Eng J* 317:242–257, 2017.
Lu J, Das S, Peters EAJF, Kuipers JAM: Direct numerical simulation of flow and mass transfer in dense fluid–particle systems with surface reactions, *Chem Eng Sci* 176:1–18, 2018.
Manger E: *Modeling and simulation of gas solid systems in curvilinear co-ordinates,* PhD thesis, Norway, 1996, Telemark Institute of Technology.
Matsen JM: Scale-up of fluidized bed processes: principle and practice, *Powder Technol* 88:237–244, 1996.
Mehrabadi M, Murphy E, Subramaniam S: Development of a gas–solid drag law for clustered particles using particle-resolved direct numerical simulation, *Chem Eng Sci* 152:199–212, 2016.

Merrow EW: Linking R&D to problems experienced in solids processing, *CEP* 81:14–22, 1985.
Mittal R, Iaccarino G: Immersed boundary methods, *Annu Rev Fluid Mech* 37:239–261, 2005.
Nieuwland JJ, van Sint Annaland M, Kuipers JAM, Swaaij v: Hydrodynamic modeling of riser reactors, *AIChE J* 42(6):1569–1582, 1996.
Patankar SV, Spalding DB: A calculation procedure for heat, mass and momentum transfer in three-dimensional parabolic flows, *Int J Heat Mass Transf* 15:1787–1806, 1972.
Peskin CS: The immersed boundary method, *Acta Numerica*: 479–517, 2002.
Pita JA, Sundaresan S: Developing flow of a gas–particle mixture in a vertical riser, *AIChE J* 39(4):541–552, 1991.
Tang Y, Peters EAJF, Kuipers JAM: A new drag correlation from fully resolved simulations of flow past monodisperse static arrays of spheres, *AIChE J* 61(2):688–698, 2015.
Tang Y, Buck B, Heinrich S, Deen NG, Kuipers JAM: Interface-resolved simulations of normal collisions of spheres on a wet surface, *AIChE Journal* 63(11):4774–4787, 2017.
Van der Hoef MA, van Sint Annaland M, Kuipers JAM: Computational fluid dynamics for dense gas–solid fluidized beds: a multi-scale modeling strategy, *Chem Eng Sci* 59:5157–5165, 2004.
Van der Hoef MA, van Sint Annaland M, Deen NG, Kuipers JAM: Numerical simulation of dense gas–solid fluidized beds: a multi-scale modeling strategy, *Annu Rev Fluid Mech* 40:47–70, 2008.
Van Sint Annaland M, Bokkers GA, Goldschmidt MJV, Olaofe OO, van der Hoef MA, Kuipers JAM: Development of a multi-fluid model for poly-disperse dense gas–solid fluidized beds, part I: model derivation and numerical implementation, *Chem Eng Sci* 64:4222–4236, 2009a.
Van Sint Annaland M, Bokkers GA, Goldschmidt MJV, Olaofe OO, van der Hoef MA, Kuipers JAM: Development of a multi-fluid model for poly-disperse dense gas solid fluidized beds, part II: segregation in binary particle mixtures, *Chem Eng Sci* 64:4237–4246, 2009b.
Verma V, Deen NG, Padding JT, Kuipers JAM: Two-fluid modelling of three-dimensional cylindrical gas–solid fluidized beds using the kinetic theory of granular flow, *Chem Eng Sci* 102:227–245, 2013.
Verma V, Padding JT, Deen NG, Kuipers JAM: Numerical investigation on the effect of pressure on fluidization in a 3D fluidized bed, *Ind Eng Chem Res* 53:17487–17498, 2014a.
Verma V, Padding JT, Deen NG, et al: Bubble dynamics in a 3D gas–solid fluidized bed using ultrafast electron beam X-ray tomography and two-fluid model, *AIChE J* 60(5): 1632–1644, 2014b.
Verma V, Padding JT, Deen NG, Kuipers JAM: Effect of bed size on hydrodynamics in 3D gas–solid fluidized beds, *AIChE J* 61(5):1492–1506, 2015.
Wen CY, Yu YH: A generalized method for predicting the minimum fluidization velocity, *AIChE J* 12(3):610–612, 1966.
Werther J: Bubble growth in large diameter fluidized beds, international fluidization conference, Pacific Grove, USA, 1975. In Keairns DL, editor: *Fluidization technology*, Washington, DC, 1976, Hemisphere Publishing Co., pp 215–235.
Yang L, Padding JT, Kuipers JAM: Modification of kinetic theory of granular flow for frictional spheres, part I: two-fluid model derivation and numerical implementation, *Chem Eng Sci* 152:767–782, 2016a.
Yang L, Padding JTJ, Kuipers JAM: Modification of kinetic theory of granular flow for frictional spheres, part II: model validation, *Chem Eng Sci* 152:783–794, 2016b.
Yang L, Padding JT, Buist KA, Kuipers JAM: Three-dimensional fluidized beds with rough spheres: validation of a two fluid model by magnetic particle tracking and discrete particle simulations, *Chem Eng Sci* 174:238–258, 2017a.
Yang L, Padding JT, Kuipers JAM: Partial slip boundary conditions for collisional granular flows at flat frictional walls, *AIChE J* 63(6):1853–1871, 2017b.

CHAPTER TWO

Advances in Coarse Discrete Particle Methods With Industrial Applications

Liqiang Lu[1], Sofiane Benyahia
National Energy Technology Laboratory, Morgantown, WV, United States
[1]Corresponding author: e-mail address: liqiang.lu@netl.doe.gov

Contents

Abstract

This chapter provides the full description of a coarse-grained discrete particle method based on a novel hard-sphere contact model for the simulation of industrial-scale fluidized bed reactors. This method is based on simple models that are easy to understand and implement in numerical codes. This technique is verified and validated for several small-scale fluidized systems where numerical data based on finer methods as well as experimental data are available. The speed of execution of this method is increased several orders of magnitudes compared to particle-based discrete methods, which allows

Advances in Chemical Engineering, Volume 53
ISSN 0065-2377
https://doi.org/10.1016/bs.ache.2017.12.001

for thousands of seconds of flow, heat, and mass transfer simulations of industrial reactors such as fluidized catalytic cracking regenerator, Methanol to Olefins reactor, and Rare Earth Elements leaching reactor, achieved in just few days using commonly available computer resources. It is now possible for the common engineer to conduct simulations of large-scale fluid–particle reactors to understand, design, and troubleshoot, as well as optimize the performance of these complex multiphase flow systems.

1. INTRODUCTION

A look at the surface of Earth from space shows that it is mostly water, with dirt covering the rest of the planet if we ignore the presence of biomass. It is, therefore, evident that granular materials in all their shapes and sizes are the second most manipulated material in industry after water (Richard et al., 2005). In fact, industrial processes transform materials from their raw state to more useful and lucrative finished products. This is true not only today but also from the earliest of human civilizations. These civilizations are usually associated with our understanding of granular materials and their transformation from the Stone Age, where stone tools and weapons were made by our earliest ancestors, to early copper/bronze metallurgy, and finally to the Iron Age. Both early and current metallurgy is based on powdery ore. We, therefore, recognize that grinding is important in the processing of mining products not only because smaller powders are easier to handle and flow, but also because of the increased surface area that makes them more reactive when further chemical processing is required. Most industries, including the chemical, energy, and pharmaceutical industries, use powders at some stage of their processes. It is, therefore, important for the chemical engineer who is involved in designing and troubleshooting these plants to understand the flow of powders. The failure to thoroughly understand the behavior of solid particles in a process can lead to severe problems in start-up and unwanted downtime of solids processing plants. It is common in the literature to point to an early Rand corporation study (Merrow et al., 1981) that found the average performance of solids processing plants was only 49% of the targeted design capacity, while 84% was achieved by plants processing only gas or liquids. A more fundamental understanding of solids handling and processing is important to better design and operate these plants. This book chapter aims at providing a predictive method that is both efficient and reliable to help design and troubleshoot plants that use solid particles.

Industry usually manipulates solid powders physically and chemically in fluidized beds because the high heat and mass transfer rates they offer allow for uniformity of temperature and products. A quote from an undergraduate chemical reactor engineering textbook perfectly summarized the high efficiency of fluidized beds: "the virtues of fluidized-bed reactor drove its competitors from the market" (Fogler, 2009). For this reason, industry has adopted the fluidized bed as the ideal fluid–solids contacting unit operation with many applications listed in the classic fluidization engineering textbook (Kunii and Levenspiel, 1991). Fluidization technology is not limited to these old applications. In fact, novel approaches to fluidization that use centrifugal force rather than just the force of gravity can open a whole new way of highly intensifying fluid–particle contacts that can occur in compact systems, which can save space and initial capital investment (De Wilde et al., 2016). Experimentally, the art of designing and scaling fluidized bed reactors is far from perfect because the hydrodynamics of fluidization change with both the system and particle sizes (Knowlton et al., 2005). This is true because of wall effects and the particle to bed diameter ratio must remain constant if all hydrodynamic scaling laws are to remain constant during scaling, which is impossible if reaction and heat transfer rates are to remain similar. Experimentation must, therefore, be augmented with mathematical modelling and numerical simulation based on computational fluid dynamics (CFD) to provide the designer with all available tools to be successful. The development and perfection of such tools, as well as their validation, are an ongoing effort. Based on this current effort, we hope that industrial practitioners contribute openly to the development and testing of new models and not be merely observers or critics of such a tremendous undertaking. This chapter offers a detailed description of a novel approach to model fluidized beds, which provides the practitioner with the ability to choose the speed and accuracy of their numerical simulation.

The continuum approach has traditionally been used to model most fluidized bed units from laboratory to industrial scales and is still currently used and actively developed by researchers and practitioners worldwide. The conservation equations of the flow of a continuum powder have been described first by Jackson's group at Princeton University and summarized in his book (Jackson, 2000), while the first simulation of a fluidized bed is attributed to Gidaspow's group at the Illinois Institute of Technology and described in his textbook on continuum kinetic theory description of fluidization (Gidaspow, 1994). It is known that powders are polydispersed in the sense that all their properties exhibit statistical distributions, such as size and density distributions. It is, therefore, difficult to find a powder with unique properties, such as size, density, or sphericity, to name just few physical properties. In some cases, a wide distribution of properties, such as particle size, may be necessary to achieve the desired flow dynamics or catalytic performance of a fluidized bed reactor as explained in the literature (Grace and Sun, 1991). It is, therefore, logical to develop continuum models based on polydispersed granular kinetic theories (Garzó et al., 2007). These theories add a heavy burden on computational resources as well as a staggering level of complexity that is intimidating to most engineers. Numerical diffusion is another unwanted effect that plagues continuum theories mostly due to discretization of convective terms especially with first-order schemes (Syamlal and Benyahia, 2013). All these issues dissipate when we model powders as what they really are: discrete particles. The discrete description of granular material allows us with a single collision model to accurately predict all flow regimes from dilute particulate flows where collisions are scarce to a static pile of grains exhibiting enduring frictional contacts. The elegance of discrete element methods (usually abbreviated in the literature as DEMs) comes from the simplicity of the mathematical models and their coding in CFD. In fact, DEM is based on simple Newton's laws of motion where the acceleration of a solid body is due to forces acting on it. The formulation of contact forces between solid particles and with wall boundaries is based either on the soft-sphere model, where the overlap between particles is modeled by spring-dashpot system, or the hard-sphere model, which is based on conservation of linear and angular momentum before/after collision. As reported in the literature, further simplifications to DEM have been conducted by coarsening these models, thus lumping many real particles into computational parcels (Lu et al., 2016a; Patankar and Joseph, 2001). Other models, such as the multiphase particle-in-cell (MP-PIC), go further by ignoring the physics of discrete particle collision (Andrews and O'Rourke, 1996). The different assumptions in coarsening DEM will be explained in more detail in this chapter, which describes a coarse discrete particle method based on hard-sphere contact model.

Adopting a coarse-grained approach to conduct fluidized bed simulations increases the speed of DEM so that these predictive models can tackle large-scale reactors of interest to most industries. DEM tracks the trajectory of every particle in a system, usually a physical three-dimensional (3D) system, and their collisions with other particles and with physical boundaries, such as the walls of a reactor. The amount of particulate material in an industrial reactor is usually expressed in tons because counting the number of particles in such huge system is neither useful nor practical. Nevertheless, it is easy to estimate that number in the trillions, which contrasts with the millions of particles usually tracked in numerical simulations by researchers in academia and industry (Xu et al., 2011). Results of one such simulation are presented in Fig. 1 showing the flow of particles in a vertical riser section of a circulating fluidized bed (CFB). The particles simulated here are polydispersed with a size ranging from about 50 to 100 μm lifted in the riser by air at ambient conditions flowing at a superficial velocity of 1 m/s with a solids mass flux of 50 kg/m^2 s. The system is hypothetical, i.e., does not exist physically, and consists of a cylinder of 0.02 m diameter and a height of 1 m with air and particles fed uniformly from the bottom and leaving freely from the open top of the riser—basically, a straight pipe without any complicated geometry. The CFD uses a uniform mesh consisting of 3.2 million cells to compute the flow of air with a particle count of about 100 million (the approximation is due to the transient nature of the flow in an open system with inlet/outlet). Fig. 1 shows a thin slice, from wall to wall, of a few particles deep in the middle of the riser. The vertical length

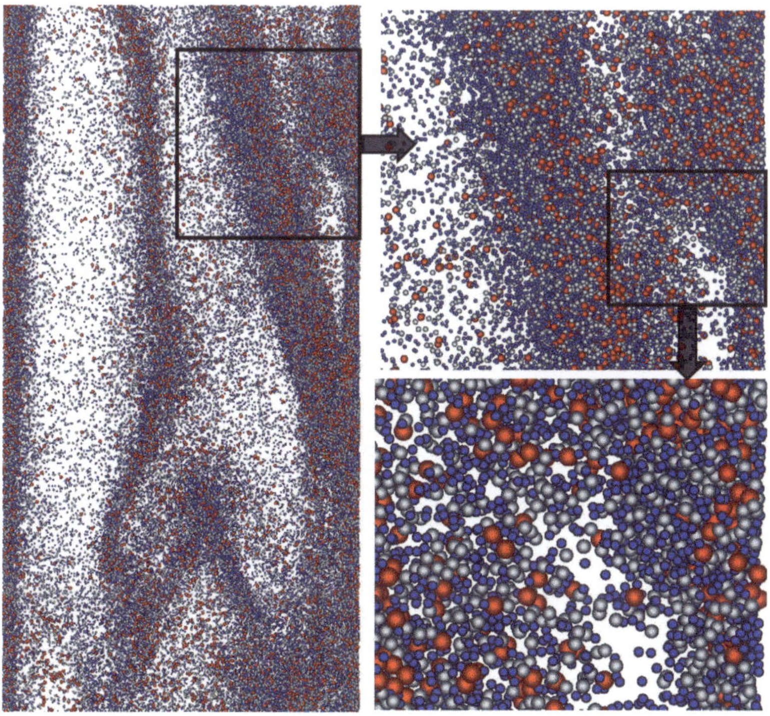

Fig. 1 Instantaneous particle position *colored* by size in a thin slice of a 3D riser with two regions of interest zoomed to show flow structure details. The size of particles is to scale with respect to the riser geometry showing the full 0.02 m width of the riser and only twice that in length centered in the middle of the riser.

of this thin slice is about twice the diameter of the riser. It shows an instantaneous picture of the flow of particles in the middle section of the riser, and demonstrates, even in this small system, the existence of clusters of particles that flow downward slowly as observed experimentally (Shaffer et al., 2013), while a more dilute and fast upward flow of gas occurs at the core of the riser. This complex motion of particles can be observed in Video 1 in the online version at https://doi.org/10.1016/bs.ache.2017.12.001 included in the supplemental material. The computation cost of conducting this numerical simulation on 256 processors is about 1 month of wall-clock time (the actual time we waited for the results) per second of simulated flow time. An industrial riser of 1 m diameter and 40 m height operating at the same condition will contain 0.1 million times more particles or 10 trillion particles. It is easy to imagine that the required resources to compute such industrial fluidized beds are prohibitive. The cost in time and resources required to conduct such simulations is a disincentive to most researchers and engineers.

The development of new predictive models requires some validation efforts to be conducted. Comparing computation results with experimental data can be fruitful to provide an idea of the predictive capabilities of the model but this is not sufficient to declare a model valid. In fact, a model is never fully valid or completely wrong. The idea is to evaluate how much uncertainty is due to major assumptions in developing any mathematical model to represent a physical reality. This physical reality is seldom well known and understood because the experimental measurements are also prone to uncertainties that can be significant. To evaluate such uncertainties, an interesting idea came from John Grace's research group at the University of British Columbia (UBC) where a "traveling" bubbling fluidized bed was designed and sent to several research groups who measured some quantities of interest under the same flow conditions for later comparing the different measurement techniques used by these research teams. The particle velocity measurements were published recently (Tebianian et al., 2016) and show some large quantitative differences even in the trends of the radial time-averaged profiles obtained with different measurement techniques (both intrusive and nonintrusive). It is clear from this sobering study that modelers who aim to obtain a perfect match with experimental data are doomed to fail. These interesting findings make it even more difficult to claim that a mathematical model that has been implemented in a CFD code is fully verified and validated (V&V). Needless to say, modelers overstepped by claiming some years back complete V&V of CFD codes, as pointed out by the same research group at UBC (Grace and Taghipour, 2004). The truth is that both experimentalists and modelers are happy to obtain correct trends of flow quantities of interest. Significant efforts will be required to obtain quantitative assessments of errors in experimental measurements and numerical simulations of fluidized beds. Even then, those assessments may only be valid for the system studied under actual flow conditions. In our current approach of attempting to accelerate discrete particle models by adopting a coarse-grained approach, the model errors are usually assessed with respect to finer models that use fewer assumptions. We still make use of experimental data when available to compare the trends of those profiles with those obtained numerically. These comparisons are also useful to motivate industrial practitioners and engineers to adopt CFD as an important tool, among other tools, for designing and troubleshooting fluidized bed reactors.

This chapter provides the complete details of a novel coarse-grained DEM approach based on an improved time-driven hard-sphere method aimed at predicting flow dynamics in fluidized bed reactors cheaply yet accurately. The details of this approach will be described and compared with other methods available to the modelers. Some aspects of V&V of this method in our CFD open-source code will be addressed by simulating small-scale fluidized beds where previously published experimental measurements and/or detailed numerical simulation results are available, then moving to pilot-scale fluidized beds where experimental and simulation data are also available, and finally large-scale industrial reactors will be simulated and compared to published industrial data. This succession of simulations from small- to large-scale reactors is aimed at attracting the attention of modelers responsible for all design stages ranging from early reactor development to final industrial-scale reactor deployment. Thus, our main objective is to prove that CFD can be a reliable partner for the engineers and practitioners who aim at designing and troubleshooting fluidized bed systems. To fulfill this purpose, we hope to deliver a model that is not limited in range or scope but puts users in the driver's seat, relying only on users to make an informed decision on how to conduct their simulation to achieve the desired level of accuracy at an acceptable speed of execution.

2. DESCRIPTION OF METHODS FOR SOLVING FLUID–PARTICLE FLOW

Discrete particle simulations of fluidized beds can be traced back to the early 1990s with the pioneering work of Tsuji et al. (1993) using CFD–DEM, Hoomans et al. (1996) using the event-driven hard sphere method (EDHS), and Ouyang and Li (1999) using the time-driven hard sphere model (TDHS). These methods avoid the continuum approach drawbacks by simply tracking the movement of individual particles and directly resolving particle–particle collisions. In CFD–DEM, the particles are assumed to be soft, and the collision is simulated by a linear spring-dashpot model. Note that in the general literature, the acronyms DEM and DPM are used interchangeably for all types of collision models, but it is reserved in this book chapter for soft-sphere collisions. This model is widely used because of its simplicity and capability to handle multibody collisions. However, it requires a much smaller time step to resolve particle–particle collisions. The EDHS handles particle–particle collisions using momentum conservation laws. This method is usually applied in dilute gas–solids flow where the mean free path between particle collisions is large. For dense phase flow, the collision frequency is much higher than that in dilute phase and the time interval between collisions is very small, rendering this method inefficient. The TDHS method solved this problem by allowing small overlaps between colliding particles; thus a larger time step can be used even for dense phase flows. However, TDHS still lacks the ability to handle multibody collisions and will result in unphysical overlaps for cases like particle packing. Lu et al. solved this problem by introducing a velocity correction term that aims to reduce the largest overlaps among colliding particles (Lu et al., 2017d). For industrial applications, computational parcels are simulated rather than real physical particles following a coarse-grained CFD–DEM or a coarse-grained hard sphere approach. These different approaches, along with MP-PIC, have been compared in recent research (Lu et al., 2017a). Ge et al. (2017) also reviewed fundamental developments and engineering applications of discrete particle simulations with a focus on multiscale structures. In this section, recent advances in these methods are summarized focusing on coarse-grained CFD–DEM and time-driven hard sphere methods.

2.1 Computational Fluid Dynamics Coupled With Soft-Sphere Contact Model

Pioneered by Tsuji et al. (1993) and Xu and Yu (1997), CFD–DEM has been widely used to investigate particle–fluid flows including hydrodynamics, heat transfer, and chemical reactions. The governing equations of this method are summarized in this section.

2.1.1 Fluid Dynamics

In CFD–DEM, the gas phase is governed by the traditional volume-averaged Navier–Stokes equations

$$\frac{\partial(\varepsilon_f\rho_f)}{\partial t}+(\nabla\cdot\varepsilon_f\rho_f\mathbf{u}_f)=0 \tag{1}$$

$$\frac{D(\varepsilon_f\rho_f\mathbf{u}_f)}{Dt}=\nabla\cdot\overline{\overline{S}}_f+\varepsilon_f\rho_f\mathbf{g}-\mathbf{I} \tag{2}$$

where ε_f is the volume fraction of fluid, ρ_f is the density of the fluid, $\mathbf{u}_f$ is the velocity of the fluid, and $\mathbf{I}$ is the drag source term. $\overline{\overline{S}}_f$ is the fluid phase stress tensor given by

$$\overline{\overline{S}}_f=-P_f\overline{\overline{I}}+\overline{\overline{\tau}}_f \tag{3}$$

where P_f is the fluid phase pressure and $\overline{\overline{\tau}}_f$ is the Newtonian fluid phase shear stress tensor

$$\overline{\overline{\tau}}_f=2\mu_f\overline{\overline{D}}_f+\lambda_f\nabla\cdot tr\left(\overline{\overline{D}}_f\right)\overline{\overline{I}} \tag{4}$$

$$\overline{\overline{D}}_f=\frac{1}{2}\left[\nabla\mathbf{u}_f+(\nabla\mathbf{u}_f)^T\right] \tag{5}$$

where $\overline{\overline{D}}_f$ is the strain rate tensor, and μ_f and λ_f are the fluid dynamic and bulk viscosities.

The interphase momentum transfer term on fluid cell c can be calculated as

$$\mathbf{I}^c=\frac{1}{v_c}\sum_{i=1}^{N_p}\frac{1}{6}\pi d_p^3\left(\nabla P_f\left(\mathbf{x}^i\right)+\frac{\beta^i}{1-\varepsilon_f}\left(\mathbf{v}_f\left(\mathbf{x}^i\right)-\mathbf{v}_p^i\right)\right)K\left(\mathbf{x}^i,\mathbf{x}_c\right) \tag{6}$$

where v_c is the volume of cell c and N_p is the number of particles influencing cell c. β^i is the drag coefficient of particle i in cell c. $\mathbf{v}_f(\mathbf{x}^i)$ is the fluid velocity interpolated at particle i and K is the interpolation weight of particle i to cell c.

2.1.2 Momentum Transfer Models

The momentum exchange between the fluid phase and particles is expressed by different drag correlations. In fluidization, the motion of a particle is mainly driven by the fluid drag force countering a constant pull of the gravitational force. Thus, accurately calculating this drag force is critical to modelling the fluidization behavior. Currently, no general drag expression can be used for all fluidization cases. The widely used drag models include empirically based models such as the Wen—Yu drag model (Wen and Yu, 1966), the Gidaspow drag model (Gidaspow, 1994), and the Syamlal–O'Brien drag model (Syamlal and O'Brien, 1987) which is based on a Richardson–Zaki type velocity–voidage correlation; drag models derived by fitting direct numerical simulation (DNS) data such as the Beetstra–Van der Hoef–Kuipers (BVK) drag model (Beetstra et al., 2007) and the Hill–Koch–Ladd drag model (Benyahia et al., 2006); and finally drag models that include the effects of unresolved mesoscale structures such as the energy minimization multiscale (EMMS) drag model (Ge and Li, 2002; Li and Kwauk, 1994; Wang and Li, 2007; Wang et al., 2008; Yang et al., 2003) and the filtered drag model (Igci et al., 2008; Radl and Sundaresan, 2014). Due to the abundance of drag correlations derived in the literature, we will only express those used directly in this study. A full discussion on the specifics of each drag correlation and its applications falls, unfortunately, outside the scope of this work and will, thus, not be covered here. A brief discussion of some of these drag models and their different applications for discrete coarse-grained models was covered in our recent publication (Lu et al., 2017b), and the reader is encouraged to pursue different strategies to understand the profound effects these drag correlations can have on simulation results. Most of these drag correlations are based on standard drag coefficient for a single particle expressed as

$$C_{\mathrm{D0}} = \begin{cases} \dfrac{24}{Re_{\mathrm{p}}}\left(1 + 0.15 Re_{\mathrm{p}}^{0.687}\right) & Re_{\mathrm{p}} < 1000 \\ 0.44 & Re_{\mathrm{p}} \geq 1000 \end{cases} \tag{7}$$

where

$$Re_{\mathrm{p}} = \frac{\varepsilon_{\mathrm{f}} \rho_{\mathrm{f}} \left|\mathbf{v}_{\mathrm{f}} - \mathbf{v}_{\mathrm{p}}\right| d_{\mathrm{p}}}{\mu_{\mathrm{f}}} \tag{8}$$

For the Wen—Yu model, the drag coefficient is calculated as

$$\beta_{\mathrm{Wen-Yu}} = \frac{3}{4} \frac{\varepsilon_{\mathrm{f}}(1-\varepsilon_{\mathrm{f}}) \rho_{\mathrm{f}} \left|\mathbf{v}_{\mathrm{f}} - \mathbf{v}_{\mathrm{p}}\right|}{d_{\mathrm{p}}} C_{\mathrm{D0}} \varepsilon_{\mathrm{f}}^{-2.65} \tag{9}$$

For the Gidaspow model, the drag coefficient is calculated as

$$\beta_{\mathrm{Gidaspow}} = \begin{cases} 150 \dfrac{(1-\varepsilon_{\mathrm{f}})^2 \mu_{\mathrm{f}}}{\varepsilon_{\mathrm{f}} d_{\mathrm{p}}^2} + 1.75 \dfrac{(1-\varepsilon_{\mathrm{f}}) \rho_{\mathrm{f}} \left|\mathbf{v}_{\mathrm{f}} - \mathbf{v}_{\mathrm{p}}\right|}{d_{\mathrm{p}}}, & \varepsilon_{\mathrm{f}} \leq 0.8 \\ \dfrac{3}{4} \dfrac{\varepsilon_{\mathrm{f}}(1-\varepsilon_{\mathrm{f}}) \rho_{\mathrm{f}} \left|\mathbf{v}_{\mathrm{f}} - \mathbf{v}_{\mathrm{p}}\right|}{d_{\mathrm{p}}} C_{\mathrm{D0}} \varepsilon_{\mathrm{f}}^{-2.65}, & \varepsilon_{\mathrm{f}} > 0.8 \end{cases} \tag{10}$$

For the Syamlal and O'Brien model, the drag coefficient is calculated as

$$\begin{aligned}
\beta_{\text{Syamlal and O'Brien}} &= \frac{3}{4}\frac{C_{so}\varepsilon_f(1-\varepsilon_f)\rho_f}{d_p}|\mathbf{v}_f-\mathbf{v}_p| \\
C_{so} &= \left(\frac{0.63}{v_r}+\frac{4.8}{\sqrt{v_r Re_p}}\right)^2 \\
v_r &= 0.5\left(A-0.06Re_p+\sqrt{(0.06Re_p)^2+0.12Re_p(2B-A)+A^2}\right) \\
A &= \varepsilon_f^{4.14} \\
B &= \begin{cases} C_2\varepsilon_f^{1.28}, & \varepsilon_f \le 0.85 \\ \varepsilon_f^{C_1}, & \varepsilon_f > 0.85 \end{cases}
\end{aligned} \tag{11}$$

For the BVK model, the drag coefficient is calculated as

$$\beta_{\text{BVK}} = \frac{18\mu\varepsilon_f(1-\varepsilon_f)F_{\text{BVK}}}{d_p^2} \tag{12}$$

$$\begin{aligned}
F_{\text{BVK}} &= \frac{10(1-\varepsilon_f)}{\varepsilon_f^2}+\varepsilon_f^2\left(1+1.5(1-\varepsilon_f)^{0.5}\right) \\
&+\frac{0.413Re_p}{24\varepsilon_f^2}\left[\frac{\varepsilon_f^{-1}+3\varepsilon_f(1-\varepsilon_f)+8.4Re_p^{-0.343}}{1+10^{3(1-\varepsilon_f)}Re_p{}^{-0.5-2(1-\varepsilon_f)}}\right]
\end{aligned} \tag{13}$$

For the EMMS model, a heterogeneous factor is introduced to modify the Wen—Yu drag model. This factor is usually a function of voidage and slip velocity calculated in every computational cell. This function is based on gas density and viscosity, particle density and size, and operating conditions such as superficial velocity and bed size. In this chapter, the EMMS drag model is used in the simulation of a methanol to olefins (MTO) reactor (Tian et al., 2015) and is expressed as follows (Lu et al., 2016b):

$$\begin{aligned}
\beta_{\text{EMMS}} &= \frac{3}{4}\frac{\varepsilon_f(1-\varepsilon_f)\rho_f|\mathbf{v}_f-\mathbf{v}_p|}{d_p}C_{D0}\varepsilon_f{}^{-2.65}H_D \\
H_D &= \begin{cases} 0.1443+\dfrac{1.6864}{1+(\varepsilon_f/0.4064)^{20.4805}}, & \varepsilon_f \le 0.4924 \\ (-0.0999+0.9586\varepsilon_f)^{1.8008}, & 0.4924<\varepsilon_f \le 0.9505 \\ (1649.4591+1648.5453\varepsilon_f)^{0.08597}, & 0.9505<\varepsilon_f \le 1 \end{cases}
\end{aligned} \tag{14}$$

The drag model for fluidized catalytic cracking (FCC) particles proposed by Gao et al. (2009) is used for the simulation of industrial FCC regenerators. This drag coefficient is expressed as

$$\beta_{\mathrm{Gao}}=\begin{cases}150\dfrac{(1-\varepsilon_{\mathrm{f}})^2\mu_{\mathrm{f}}}{\varepsilon_{\mathrm{f}}d_{\mathrm{p^*}}^2}+1.75\dfrac{(1-\varepsilon_{\mathrm{f}})\rho_{\mathrm{f}}|\mathbf{v}_{\mathrm{f}}-\mathbf{v}_{\mathrm{p}}|}{d_{\mathrm{p^*}}}, & \varepsilon_{\mathrm{f}}\le 0.800\\ \dfrac{5}{72}\dfrac{\varepsilon_{\mathrm{f}}(1-\varepsilon_{\mathrm{f}})\rho_{\mathrm{f}}|\mathbf{v}_{\mathrm{f}}-\mathbf{v}_{\mathrm{p}}|}{d_{\mathrm{p^*}}(1-\varepsilon_{\mathrm{f}})^{0.293}}C_{\mathrm{D^*}}, & 0.800<\varepsilon_{\mathrm{f}}\le 0.933\\ \dfrac{3}{4}\dfrac{\varepsilon_{\mathrm{f}}(1-\varepsilon_{\mathrm{f}})\rho_{\mathrm{f}}|\mathbf{v}_{\mathrm{f}}-\mathbf{v}_{\mathrm{p}}|}{d_{\mathrm{p}}}C_{\mathrm{D0}}{\varepsilon_{\mathrm{f}}}^{-2.65}, & 0.933<\varepsilon_{\mathrm{f}}\le 0.990\\ \dfrac{3}{4}\dfrac{\varepsilon_{\mathrm{f}}(1-\varepsilon_{\mathrm{f}})\rho_{\mathrm{f}}|\mathbf{v}_{\mathrm{f}}-\mathbf{v}_{\mathrm{p}}|}{d_{\mathrm{p}}}C_{\mathrm{D0}}, & 0.990<\varepsilon_{\mathrm{f}}\le 1.000\end{cases}\tag{15}$$

where

$$C_{\mathrm{D^*}}=\begin{cases}\dfrac{24}{Re_{\mathrm{p^*}}}\left(1+0.15{Re_{\mathrm{p^*}}}^{0.687}\right) & Re_{\mathrm{p^*}}<1000\\ 0.44 & Re_{\mathrm{p^*}}\ge 1000\end{cases}\tag{16}$$

$$Re_{\mathrm{p^*}}=\frac{\varepsilon_{\mathrm{f}}\rho_{\mathrm{f}}\left|\mathbf{v}_{\mathrm{f}}-\mathbf{v}_{\mathrm{p}}\right|d_{\mathrm{p^*}}}{\mu_{\mathrm{f}}}\tag{17}$$

2.1.3 Solid Particle Dynamics

The Lagrangian method is used to track the motion of particles. For each solids particle, the momentum equation takes the familiar form of Newton's law of motion

$$m_{\mathrm{p}}\frac{d\mathbf{v}_{\mathrm{p}}}{dt}=m_{\mathrm{p}}\mathbf{g}-\frac{\pi}{6}d_{\mathrm{p}}^3\nabla P_{\mathrm{f}}+\frac{\beta\left(\mathbf{v}_{\mathrm{f}}(\mathbf{x}^i)-\mathbf{v}_{\mathrm{p}}\right)}{1-\varepsilon_{\mathrm{f}}}\frac{\pi}{6}d_{\mathrm{p}}^3+\mathbf{F}_{\mathrm{c}}\tag{18}$$

$$I\frac{d\mathbf{w}_{\mathrm{p}}}{dt}=\mathbf{T}\tag{19}$$

where m_{p} is the mass of the particle and d_{p} its diameter. $\mathbf{v}_{\mathrm{p}}$ is translation velocity of particle and $\mathbf{w}_{\mathrm{p}}$ is its angular velocity. The moment of inertia I is calculated by $m_{\mathrm{p}}d_{\mathrm{p}}^2/10$. On the right-hand side, the forces considered include the gravity force, pressure gradient force, drag force, and the contact force ($\mathbf{F}_{\mathrm{c}}$). The contact force is calculated by the DEM in normal ($\mathbf{F}^{\mathrm{n}}$) and tangential ($\mathbf{F}^{\mathrm{t}}$) directions and is described below.

The normal contact force acting on particle i due to its collision with particle j is calculated as

$$\mathbf{F}_{ij}^{\mathrm{n}}=\left(-k_{\mathrm{n}}\delta_{\mathrm{n}}+\eta_{\mathrm{n}}\dot{\delta}_{\mathrm{n}}\right)\mathbf{n}_{ij}\tag{20}$$

where k_n is the spring stiffness, δ_n is the thickness of the overlapped region, and the dot represents time derivative, $\mathbf{n}_{ij}$ is a unit vector along the line of centers from j to i, and η_n is the damping coefficient defined as

$$\eta_n = \sqrt{2k_n m_p}\frac{\ln(1/e_p)}{\sqrt{\pi^2 + \left[\ln(1/e_p)\right]^2}} \tag{21}$$

where e_P denotes the restitution coefficient.

The tangential contact force is calculated similarly to the normal forces as

$$\mathbf{F}_{ij}^{t} = \left(-k_t\delta_t + \upsilon_t\dot{\delta}_t\right)\mathbf{t}_{ij} \tag{22}$$

where δ_t is the tangential displacement. It is caused by the relative movement of colliding particles in tangential direction $\mathbf{t}_{ij}$, which can be calculated as

$$\mathbf{t}_{ij} = \frac{\mathbf{v}_{t,\,ij}}{\left|\mathbf{v}_{t,\,ij}\right|} \tag{23}$$

Where $\mathbf{v}_{t,ij}$ is the tangential relative velocity. It can be calculated as

$$\mathbf{v}_{t,\,ij} = \left(\mathbf{v}_i - \mathbf{v}_j\right) - \left(\mathbf{v}_i - \mathbf{v}_j\right)\cdot \mathbf{n}_{ij}\mathbf{n}_{ij} \tag{24}$$

The tangential force is also limited by the finite Coulomb friction between colliding particles with a sliding friction coefficient of μ. Thus, it is defined as

$$\mathbf{F}_{ij}^{t} = \min\left(\mathbf{F}_{ij}^{t}, \mu\mathbf{F}_{ij}^{n}\right) \tag{25}$$

Finally, the collision force on the tracked i particle is

$$\mathbf{F}_c = \sum_{j=1,\,j\neq i}^{N}\left(\mathbf{F}_{ij}^{n} + \mathbf{F}_{ij}^{t}\right) \tag{26}$$

As for particle rotation, the torque on each tracked i particle is calculated as

$$\mathbf{T} = \sum_{j=1,\,j\neq i}^{N}\left(L\mathbf{n} \times \mathbf{F}_{ij}^{t}\right) \tag{27}$$

where L is the distance from particle center to contact point.

2.1.4 Heat Transfer Model

Generally, the heat transfer mechanisms include convection, conduction, and radiation. In fluidized beds, the main heat transfer mechanisms for particles include particle–fluid convection, particle–fluid–particle conduction, particle–fluid–wall conduction, particle–particle contact conduction, and particle–wall contact conduction. Radiative heat transfer is not treated here because it is not of interest to this research and is, thus, deferred for future development. Thus, the conservation equation for heat transfer of particle i can be written as

$$m_i c_{\mathrm{p},i} \frac{dT_i}{dt} = Q_{i,\mathrm{conv}} + Q_{i,\mathrm{w}} + \sum_{j=1}^{K_i} Q_{i,j} \tag{28}$$

Where $Q_{i,\mathrm{conv}}$ is particle–fluid convection. $Q_{i,\mathrm{w}}$ is particle–wall conduction, which includes particle–fluid–wall conduction and particle–wall contact conduction. $Q_{i,j}$ is particle–particle conduction, which includes particle–fluid–particle conduction and particle–particle contact conduction.

The particle–fluid convection is proportional to temperature difference, surface area, and the convective heat transfer coefficient. This convection is calculated by

$$Q_{i,\mathrm{conv}} = h_{i,\mathrm{conv}} \pi d_{\mathrm{p}}^2 (T_{\mathrm{f}} - T_i) \tag{29}$$

where the convective heat transfer coefficient $h_{i,\mathrm{conv}}$ is a function of the particle Nusselt number, and the correlation given by Gunn (1978) is used

$$\begin{aligned} Nu_i = & \left(7 - 10\varepsilon_{\mathrm{f}} + 5\varepsilon_{\mathrm{f}}^2\right)\left(1.0 + 0.7 Re_i^{0.2} Pr^{0.33}\right) \\ & + \left(1.33 - 2.4\varepsilon_{\mathrm{f}} + 1.2\varepsilon_{\mathrm{f}}^2\right) Re_i^{0.7} Pr^{0.33} \end{aligned} \tag{30}$$

$$Nu_i = \frac{h_{i,\mathrm{conv}} d_{\mathrm{p}}}{k_{\mathrm{f}}}, \quad Pr_i = \frac{\mu_{\mathrm{f}} C_{\mathrm{f}}}{k_{\mathrm{f}}} \tag{31}$$

where d_{p} and k_{f} are the particle diameter and fluid thermal conductivity. Re_i is the local Reynolds number of particle i. C_{f} is the specific heat of fluid.

As shown in the following equation, particle–wall conduction includes particle–fluid–wall conduction and particle–wall collision conduction.

$$Q_{i,\mathrm{w}} = Q_{i,\mathrm{w}}^{\mathrm{pfw}} + Q_{i,\mathrm{w}}^{\mathrm{pwc}} \tag{32}$$

For particle–fluid–wall conduction, the model introduced by Rong and Horio (1999) is used in this research, as shown in Fig. 2. They hypothesize that each particle is surrounded by a gas layer with a constant thickness, r_{len}; particle–fluid–wall heat transfer initiates when the gas layer intersects

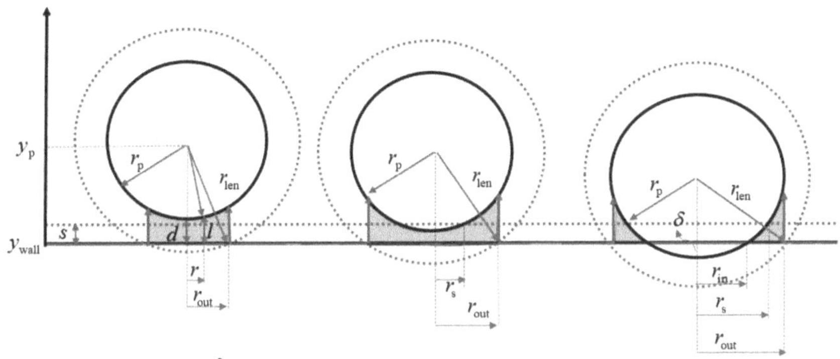

Fig. 2 Particle–fluid–wall heat transfer model. From (A) to (C), the particle is getting closer to the wall. In case (A) only the gas layer contacts the wall and the distance is larger than *s*; in case (B) the distance is smaller than *s*; in case (C) the particle is in contact with wall. *Modified from Lu L, Morris A, Li T, Benyahia S: Extension of a coarse grained particle method to simulate heat transfer in fluidized beds.* Int J Heat Mass Transf *111:723–735, 2017c with permission for the reprinted content. Copyright (2017) Elsevier.*

with the wall; heat transfer occurs across paths parallel to the axis joining the particle's center and the wall; a uniform gas layer, with a constant thickness, *s*, separates the particle and the wall; the temperature is assumed to be uniform inside a particle. Thus, this term can be calculated as

$$Q_{i,\mathrm{w}}^{\mathrm{pfw}} = 2\pi k_{\mathrm{g}}(T_{\mathrm{w}} - T_{\mathrm{i}}) \int_{r_{\mathrm{in}}}^{r_{\mathrm{out}}} \frac{r}{l} dr \tag{33}$$

where k_g is the gas thermal conductive coefficient, T_w and T_i are the temperatures of wall and particle, l is the distance between particle and wall, and the integration from r_{in} to r_{out} delineates the region of particle–fluid–wall conduction.

For the particle–wall contact conduction term, the model proposed by Batchelor and Brien (1977) is adopted here. When a particle contacts the wall, the conduction heat transfer can be calculated as

$$Q_{i,\mathrm{w}}^{\mathrm{pwc}} = \frac{4r_{\mathrm{c}}(T_{\mathrm{w}} - T_i)}{(1/k_{\mathrm{pi}} + 1/k_{\mathrm{w}})} \tag{34}$$

where r_c is the contact radius, k_{pi} and k_w are thermal conductive coefficients of the particle and the wall, T_w and T_i are the temperatures of the wall and the particle, and r_c can be calculated by the following equation:

$$r_{\mathrm{c}} = \sqrt{r_{\mathrm{p}}^2 - (r_{\mathrm{p}} - \delta)^2} = \sqrt{2r_{\mathrm{p}}\delta - \delta^2} \tag{35}$$

Finally, for the fluid phase, the conservation equation of thermal energy is expressed as

$$\varepsilon_f \rho_f C_{pf} \left(\frac{\partial T_f}{\partial t} + \mathbf{u}_f \cdot \nabla T_f \right) = -\nabla \cdot \vec{q}_f + S_{pf} + S \tag{36}$$

where S_{pf} is the total heat transfer from solid particles and S is any other source terms, such as heat generated or consumed due to chemical reactions. The conductive heat flux is described by Fourier's law

$$\vec{q}_f = -\varepsilon_f k_f \nabla T_f \tag{37}$$

$$S_{pf} = -\frac{1}{V_{cell}} \sum_{i=1}^{N_P} Q_{i,conv} \tag{38}$$

2.1.5 Species Transport With Chemical Reactions

For Euler–Lagrange simulations, heterogeneous chemical reactions can be simulated with particle-scale chemical reaction kinetic models, such as the shrinking core model, where the reaction rate is a function of the particle diameter. In MFIX-DEM, all chemical reaction rates are defined by the user through the modification of a file named usr_des_rrates.f. In the discrete particle model, the mass, density, species fraction, and radius of each particle are tracked. The particle's mass variation due to chemical reaction is given by

$$\frac{dm_p}{dt} = \sum_{n=1}^{N_s} R_{sn} \tag{39}$$

where R_{sn} is the rate of production (or consumption) of species n in this particle and N_s is the number of species. m_p is the particle mass and it equals:

$$m_p = \sum_{n=1}^{N_s} X_{pn} m_p \tag{40}$$

The mass variation of species n due to chemical reaction is given by

$$\frac{d(X_{pn} m_p)}{dt} = R_{sn} \tag{41}$$

Using the product rule and Eq. (40), the mass fraction of species n in this particle (X_{pn}) can be further expressed as

$$\frac{dX_{pn}}{dt} = \frac{1}{m_p} \left(R_{sn} - X_{pn} \sum_{n=1}^{N_s} R_{sn} \right) \tag{42}$$

Since the mass of particles changes due to chemical reactions, either the density or the diameter of this particle should be modified. If the density of the particle is assumed to be constant by the user, then the diameter of the particle is recalculated according to

$$d_\mathrm{p} = \left(\frac{6m_\mathrm{p}}{\pi\rho_\mathrm{p}}\right)^{1/3} \tag{43}$$

On the other hand, if the particle diameter is assumed to be constant, then the density of the particle is recalculated according to

$$\rho_\mathrm{p} = \frac{6m_\mathrm{p}}{\pi d_\mathrm{p}^3} \tag{44}$$

For species fraction in the fluid phase, the following continuum transport equation is solved:

$$\frac{\partial(\varepsilon_\mathrm{f}\rho_\mathrm{f}X_n)}{\partial t} + \nabla\cdot(\varepsilon_\mathrm{f}\rho_\mathrm{f}\mathbf{u}_\mathrm{f}X_n) = \nabla\cdot(D_n\nabla X_n) + R_{\mathrm{f}n} \tag{45}$$

where X_n is the mass fraction of fluid species n, D_n its diffusion coefficient, and $R_{\mathrm{f}n}$ is the mass transfer of species n between the particles and the fluid phase, and is calculated as

$$R_{\mathrm{f}n} = \frac{1}{V_\mathrm{cell}}\sum_{i=1}^{N_\mathrm{k}} R_{\mathrm{f}n}^i \tag{46}$$

where V_cell is the volume of a CFD cell, N_k is the number of particles in this cell, $R_{\mathrm{f}n}^i$ is the rate of fluid phase species production (or consumption) due to heterogeneous reactions occurring between particle i and the fluid phase.

2.2 Coarse-Grained Soft-Sphere Contact Model

To reduce computation costs, as discussed in the introduction, the combination of the CFD–DEM method with the parcel-based method results in the coarse-grained CFD–DEM. Only computational spheres, or parcels, are tracked because all particles within a parcel follow the exact same trajectory and have identical properties that may change in time due to heat/mass transfer. Sakano et al. (2000) simulated a bubbling fluidized bed in which many physical particles are lumped into a computational sphere, and obtained good agreement of bubble diameters by assuming that the particles in the computational parcel are in a face-centered cubic arrangement.

Around the same time, Patankar and Joseph (2001) investigated the accuracy of the parcel-based method by simulating a simple sedimentation case, a gas–solids bubbling fluidized bed, and a liquid–solids fluidized bed. The predicted sedimentation results compared well with experimental data, and for bubbling fluidized bed a similar bubbling pattern is observed when using two different parcel sizes. These pioneering studies gained the attention of researchers because increasing computer speeds allows this technique to be used to investigate pilot and even industrial-scale reactors. In this section, the basic assumptions and governing equations of coarse-grained CFD–DEM are summarized, followed by discussion of recent advances in this area and extension to heat transfer.

2.2.1 Basic Assumptions in Coarse-Grained Soft-Sphere Contact Model

In the coarse-grained CFD–DEM model, the averaged Navier–Stokes equations for the continuum fluid phase are solved in the same way as in traditional CFD–DEM with the notable exception that the interphase momentum ($\mathbf{I}^{CG}$), heat (S^{CG}), and mass transfers (R^{CG}) are recalculated based on the number of real particles in a computational parcel. Fig. 3 shows some basic assumptions in coarse-grained CFD–DEM where the main idea behind coarse graining is the lumping of real particles into a coarse-grained or computational parcel (CGP). The number of real particles in a CGP is represented by a statistic weight W. In this figure, we take a computational cell in an original system (physical system made of real particles) as an example to explain the assumptions and calculation method for drag force (a), heat transfer (b), chemical reactions (c), collision forces (d), and collision torques (e).

For coarse-grained CFD–DEM the fluid phase is resolved at grid scale, while the solids phase is resolved at parcel scale in the same way as in traditional CFD–DEM. All the particles in a computational cell share the same (or similar if using interpolated values) voidage, fluid velocity, temperature, and species fractions. The first major assumption in coarse-grained CFD–DEM is that all the real particles in a CGP are subjected to the same drag force. The error in this assumption is related to the number of real particles in the parcel, and to the gradient of Eulerian field values such as voidage and fluid velocity. This means that for a homogeneous flow, a much larger W (recall that W stands for number of real particles in a CGP) can be used, while for heterogeneous flows characterized by bubbles and clusters, either a smaller W or further corrections to drag force should be used. According

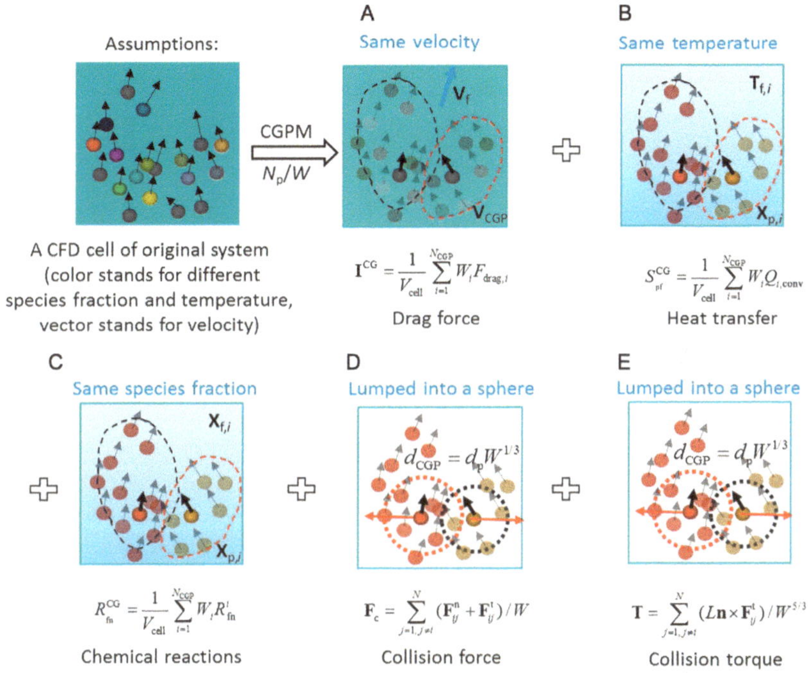

Fig. 3 Basic assumptions in coarse-grained CFD–DEM. N_p stands for number of real particles in original system, W is the statistic weight stands for number of real particles in a coarse-grained particle, N_{CGP} is number of coarse-grained particles.

to this assumption, the momentum exchange between parcels and the fluid phase is calculated as

$$\mathbf{I}^{CG} = \frac{1}{V_{cell}} \sum_{i=1}^{N_{CGP}} W_i F_{drag,i} \tag{47}$$

The second assumption in coarse-grained CFD–DEM is that all the real particles in a CGP have the same temperature. The error in this assumption is very small if the dominating heat transfer mechanism is fluid–particle convection, which is a reasonable assumption because fluidized beds are highly efficient in convecting heat since temperature gradients in the fluid phase are usually very small. Thus, the heat transfer between fluid phase and parcels is calculated as

$$S_{pf}^{CG} = \frac{1}{V_{cell}} \sum_{i=1}^{N_{CGP}} W_i Q_{i,conv} \tag{48}$$

The third assumption in coarse-grained CFD–DEM is that all the real particles in a CGP have the same species mass fractions. The error in this assumption is similar to the previous two assumptions since species transport occurs in a similar manner as heat and momentum transport. Thus, the mass transfer between fluid and parcels is calculated as

$$R_{\mathrm{fn}}^{\mathrm{CG}} = \frac{1}{V_{\mathrm{cell}}} \sum_{i=1}^{N_{\mathrm{CGP}}} W_i R_{\mathrm{fn}}^{i} \tag{49}$$

The fourth assumption in coarse-grained CFD–DEM is that all real particles in a CGP are subjected to the same collision force. This is a major assumption since the collision between particles inside the CGPs is skipped. To calculate the collision force, an imaginary sphere of same density as real particles is constructed, the diameter of which can be calculated as

$$d_{\mathrm{CGP}} = d_{\mathrm{p}} W^{1/3} \tag{50}$$

Then, the collision force on the parcels is calculated the same as in DEM and divided by statistic weight to obtain the collision force acting on a single particle in the parcel.

$$\mathbf{F}_{\mathrm{c}} = \sum_{j=1, j \neq i}^{N_{\mathrm{CGP}}} \left(\mathbf{F}_{ij}^{\mathrm{n}} + \mathbf{F}_{ij}^{\mathrm{t}} \right) / W \tag{51}$$

The fifth assumption in coarse-grained CFD–DEM is that all real particles in a CGP are subjected to the same collision torque. This follows the fourth assumption exactly, and the corresponding torque is calculated as

$$\mathbf{T} = \sum_{j=1, j \neq i}^{N_{\mathrm{CGP}}} \left(L\mathbf{n} \times \mathbf{F}_{ij}^{\mathrm{t}} \right) / W^{5/3} \tag{52}$$

2.2.2 Recent Advances in Coarse-Grained Soft-Sphere Contact Model

Apart from the five basic assumptions expressed in the above section, there are two important considerations for coarse-grained CFD–DEM simulation: the size and collision parameters of CGP.

The selection of a parcel size is important because it determines the simulation performances in terms of accuracy and speed. The speed of a simulation is affected by computational resources that can be dedicated to an application. Some researchers in academia may be able to wait for months for simulation results, while an industrial researcher may require faster results. The accuracy of a simulation is affected by the ratio of parcel size to particle diameter. When this ratio is equal to unity, the simulation becomes DEM-based achieving results that are of the highest possible accuracy. As this ratio increases, the speed of the simulation increases drastically but its accuracy deteriorates. Apart from an increase in speed, general criteria for selecting a value for this size ratio is not yet available. However, for systems with distinct mesoscale structures, like bubbles and clusters, the parcel size should be small enough to resolve the deformation, aggregation, and breakage of bubbles or clusters (Lu et al., 2014, 2017b). The size of bubbles and clusters can be estimated from empirical corrections or the EMMS model (Lu et al., 2014; Shi et al., 2011; Wang et al., 2008). To quantify model uncertainties due to coarsening of particles, the most accurate solution is first sought, then the uncertainty is easily determined by comparison. As an example, in two-fluid models (TFM) where the solids phase is treated as a continuum, the most accurate simulation results are obtained for a grid-independent solution. For the coarse-grained method, a "grid-independent" solution corresponds to a DEM solution where the computational parcel size is reduced to that of an individual physical particle. In small fluidized beds where the number of particles is not excessive, it is possible to quantify exactly the uncertainty associated with coarse graining. However, in large or industrial fluidized beds where the particles are quantified in tons rather than numbers, it is currently impossible to conduct DEM simulations to obtain a grid-independent solution. In this case, we propose to quantify uncertainties in fluidized beds of smaller scales, then extrapolate to larger systems to estimate the errors associated with coarsening. Such a detailed study is yet to be conducted and is left for future work. We anticipate that these errors will depend on the physics driving a particular system: for particle–particle dominated interactions (such as dense granular flows) the errors may be large, while these errors are expected to be smaller for cases where the fluid–particle drag force is dominant. In summary, the size of CGP should be larger than the particle diameter and smaller than the minimum grid-scale structures like CFD cell size, bubble size, and cluster size (Lu et al., 2014).

$$d_{\mathrm{p}} \leq d_{\mathrm{CGP}} \leq \min\left(d_{\mathrm{cl,min}}, d_{\mathrm{b,min}}, d_{\mathrm{cell,min}}\right) \tag{53}$$

The collision parameters of CGP directly influence the particle–particle and particle–wall collision forces and can influence the simulation accuracy, especially in dense phase flow where collision is the dominate mechanism for particle transport. In linear spring-dashpot force model, the main collision parameters include spring constant, restitution coefficient, and friction coefficient. The same values for these three parameters were used in the pioneering work of Sakano et al. (2000), and also by Patankar and Joseph (2001). Recently, different researchers developed a variety of scaling laws to scale these parameters from particle to parcel size based on different criteria.

Sakai and Koshizuka (2009) assumed the overlap and collision force between colliding parcels and the original particles to be the same. This implies that when two CGPs collide, all the real particles in one CGP are subject to binary collisions with the particles from the other colliding CGP. Thus, the spring coefficient and damping coefficient must be scaled by W times in this model. Since the friction coefficient is directly related to the repose angle, the same value is used for CGP. Thus, the three parameters are scaled according to

$$k_{\mathrm{n,CGP}} = W k_{\mathrm{n,p}} \tag{54}$$

$$e_{\mathrm{CGP}} = e_{\mathrm{p}} \tag{55}$$

$$\mu_{\mathrm{CGP}} = \mu_{\mathrm{p}} \tag{56}$$

Benyahia and Galvin (2010) estimated the errors of this method for shear flow, bubbling fluidized bed, and vertical riser flow. They proposed a modification to the restitution coefficient based on the fact that collisions between particles within the parcel are ignored. The restitution coefficient is modified as

$$\frac{\ln(e_{\mathrm{CGP}})}{\ln(e_{\mathrm{p}})} = \sqrt{W} \tag{57}$$

Radl et al. (2011) proposed a different scaling law based on equal energy densities in the original and the coarse-grained system. This means that the density of the particles, as well as the translational velocity must be invariant. From the dimensionless form of the mass-spring-dashpot equation, they derived that the spring constant should be scaled by $W^{1/3}$ times. Thakur et al. (2016) also validated this relationship by simulating the loading process in a frictionless granular system. This proposed scaling law is written as

$$k_{\mathrm{n,CGP}} = W^{1/3} k_{\mathrm{n,p}} \tag{58}$$

$$e_{\mathrm{CGP}} = e_{\mathrm{p}} \tag{59}$$

$$\mu_{\mathrm{CGP}} = \mu_{\mathrm{p}} \tag{60}$$

Lu et al. (2014) and Lu et al. (2016c) proposed a coarse-grained CFD–DEM model along with EMMS-DPM. In this method, the local EMMS model (Wang and Li, 2007) is first used to calculate the distribution of cluster size or bubble sizes in the fluidized bed. Then, the parcel size is set to be smaller than most of the clusters or bubbles, and the CFD grid size is set to be 2–5 times the parcel size. Since there can be hundreds or thousands of real particles in a computational parcel, the EMMS drag model is also used to calculate drag forces. The process of lumping particles together reduces the collision frequency, which directly influences the energy dissipation. To account for this error, an effective restitution coefficient was proposed, based on kinetic theory of granular flow, by assuming the energy dissipation during collisions for the original system and the coarse-grained system are identical. They also found that the spring stiffness has little influence on simulation results of the CFB riser. Finally, in their model, these three parameters are scaled as

$$k_{n,CGP} = k_{n,p} \tag{61}$$

$$e_{CGP} = \sqrt{1 + \left(e_p^2 - 1\right) W^{1/3}} \tag{62}$$

$$\mu_{CGP} = \mu_p \tag{63}$$

They simulated a bubbling fluidized bed, a CFB riser, and the full loop of a CFB. The simulated bed height, pressure drop profile, solid velocity profile, and solid volume fractions compared well with experiment data and DEM results.

2.2.3 Extension to Heat Transfer

Convective heat transfer between particles/parcels and the fluid, the details of which were previously explained, is the most important mechanism of heat transfer in fluidized bed systems. For general multiphase flows, however, particle–fluid–wall heat conduction can also play an important role if the walls of the reactor possess high heat transfer efficiency such as a stirred tank with a cooling jacket. In coarse-grained CFD–DEM, the collision between particle and wall is calculated using d_{CGP} as the collision diameter, and the parcel to wall distance l will be different from the original system when a parcel collides with the wall. To ensure the particle–fluid–wall heat transfer is identical between the CGP system and the original system, a nondimensional form of particle–fluid–wall heat transfer for CGP is derived using particle diameter as the reference length (Lu et al., 2017c)

$$\overline{Q}_{i,\mathrm{w}}^{\mathrm{pfw}} = \frac{Q_{i,\mathrm{w}}^{\mathrm{pfw}}}{2\pi k_{\mathrm{g}}(T_{\mathrm{w}} - T_i)r_{\mathrm{p}}} = \int_{\overline{r}_{\mathrm{in}}}^{\overline{r}_{\mathrm{out}}} \frac{\overline{r}}{\overline{l}} d\overline{r} \tag{64}$$

The distance l will decrease as the particle moves closer to the wall, and when the particle is in contact with the wall, or $l=0$, a singularity will appear in this equation. This singularity is solved by assuming a minimum distance, s, between particle and wall. Thus, the integral can be converted to three nonsingular functions according to the particle–wall distance

$$\overline{Q}_{i,\mathrm{w}}^{\mathrm{pfw}} = \int_{\overline{r}_{\mathrm{in}}}^{\overline{r}_{\mathrm{out}}} \frac{\overline{r}}{\overline{l}} d\overline{r} = \begin{cases} \int_{0}^{\overline{r}_{\mathrm{out}}} \frac{\overline{r}}{\overline{l}} d\overline{r} & \overline{s} < \overline{d} < \overline{r}_{\mathrm{len}} \\ \int_{0}^{\overline{r}_{\mathrm{s}}} \frac{\overline{r}}{\overline{s}} d\overline{r} + \int_{\overline{r}_{\mathrm{in}}}^{\overline{r}_{\mathrm{out}}} \frac{\overline{r}}{\overline{l}} d\overline{r} & 0 < \overline{d} < \overline{s} \\ \int_{\overline{r}_{\mathrm{in}}}^{\overline{r}_{\mathrm{s}}} \frac{\overline{r}}{\overline{s}} d\overline{r} + \int_{\overline{r}_{\mathrm{s}}}^{\overline{r}_{\mathrm{out}}} \frac{\overline{r}}{\overline{l}} d\overline{r} & \overline{d} < 0 \end{cases} \tag{65}$$

For brevity, the integration results and the geometrical relationships among r_s, r_{in}, r_{out}, l, and r are not reproduced here since these details are available elsewhere (Lu et al., 2017c). The nondimensional form of particle–fluid–wall heat transfer for the original system and coarse-grained system can be written as

$$\overline{Q}_{i,\mathrm{w}}^{\mathrm{pfw,rp}} = f\left(\overline{\delta}_{\mathrm{rp}}, \overline{r}_{\mathrm{len,rp}}, \overline{s}_{\mathrm{rp}}\right) \tag{66}$$

$$\overline{Q}_{i,\mathrm{w}}^{\mathrm{pfw,CGP}} = f\left(\overline{\delta}_{\mathrm{CGP}}, \overline{r}_{\mathrm{len,CGP}}, \overline{s}_{\mathrm{CGP}}\right) \tag{67}$$

Therefore, the nondimensional particle–fluid–wall heat transfer is only related to the nondimensional particle wall distance, gas layer, and particle wall minimum distance. To ensure the heat conduction is the same for both systems, we must assume the nondimensional minimum distance and gas layer length are the same for the original system and coarse-grained system

$$\overline{r}_{\mathrm{len,CGP}} = \overline{r}_{\mathrm{len,rp}}, \quad \overline{s}_{\mathrm{CGP}} = \overline{s}_{\mathrm{rp}} \tag{68}$$

This relationship is also expressed graphically in Fig. 4. It can be seen that when the particle is not in contact with the wall ($\delta_{\mathrm{n}} < 0$), the heat transfer increases as the particle–wall distance decreases. However, when the particle contacts the wall, the heat transfer remains almost constant. Thus, we only need to ensure the same nondimensional particle–wall distance before a particle contacts the wall.

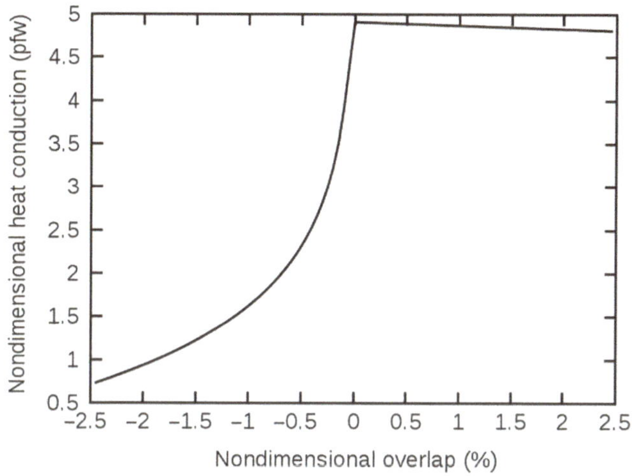

Fig. 4 Nondimensional particle–fluid–wall heat transfer dependence on nondimensional overlap (nondimensional gas layer is set to 1.04 and nondimensional minimum distance is set to 0.001).

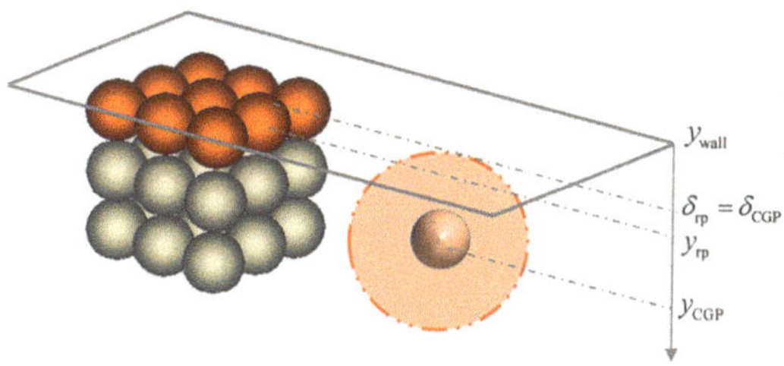

Fig. 5 Relationship of wall distance for original and coarse-grained systems. For the original system only k^2 particles (*top layer orange-colored* particles) are performing particle–fluid–wall heat transfer. *Modified from Lu L, Morris A, Li T, Benyahia S: Extension of a coarse grained particle method to simulate heat transfer in fluidized beds.* Int J Heat Mass Transf *111:723–735, 2017c with permission for the reprinted content. Copyright (2017) Elsevier.*

For dimensionless particle wall distance, as shown in Fig. 5, we assume the particle center of the coarse-grained system to be identical to the averaged center of the original system. Thus, the distance to the wall is the same for both the original system and the coarse-grained system.

The CGP diameter d_{CGP} is used for both collision dynamics and particle–fluid–wall heat transfer. Thus, the relationship of dimensionless particle wall distance in the original system and the coarse-grained system is

$$\overline{\delta}_{rp} = k\overline{\delta}_{CGP} \tag{69}$$

Then, the dimensionless form particle–fluid–wall heat transfer for the coarse-grained system can be written as

$$\overline{Q}_{i,\mathrm{w}}^{\mathrm{pfw,CGP}} = f\left(k\overline{\delta}_{\mathrm{CGP}}, \overline{r}_{\mathrm{len,CGP}}, \overline{s}_{\mathrm{CGP}}\right) \tag{70}$$

For the original system, only k^2 particles (top layer orange colored particles in Fig. 5) are performing particle–fluid–wall heat transfer. Thus, the average value should be

$$\overline{Q}_{\mathrm{avg}}^{\mathrm{pfw,rp}} = \sum_{i=1}^{W} \overline{Q}_{i,\mathrm{w}}^{\mathrm{pfw,rp}} / W = \overline{Q}_{i,\mathrm{w}}^{\mathrm{pfw,rp}} / k \tag{71}$$

Considering $\overline{Q}_{i,\mathrm{w}}^{\mathrm{pfw,rp}} = \overline{Q}_{i,\mathrm{w}}^{\mathrm{pfw,CGP}}$ and Eq. (60), we can get

$$Q_{\mathrm{avg}}^{\mathrm{pfw,rp}} = Q_{\mathrm{avg}}^{\mathrm{pfw,CGP}} / k^2 \tag{72}$$

This equation reveals that the particle–fluid–wall heat conduction in CGP method (CGPM) is k^2 times larger than in the original system. To make sure that the two systems give the same result, the calculated particle–fluid–wall heat conduction in CGPM must be rescaled by a factor of $1/k^2$.

According to recent research (Zhou et al., 2009), in fluidized beds, particle–particle conduction through the contacting area accounts for less than 3% of the total conduction term. From an engineering point of view, this term may be neglected. However, for completeness, this conductive heat transfer model is described in Eqs. (73)–(80).

For the original system and the coarse-grained system, using Eq. (35), the colliding radius r_c can be written as

$$r_{\mathrm{c}}^{\mathrm{rp}} = \sqrt{2r_{\mathrm{rp}}\delta_{\mathrm{rp}} - \delta_{\mathrm{rp}}^2} \tag{73}$$

$$r_{\mathrm{c}}^{\mathrm{CGP}} = \sqrt{2r_{\mathrm{CGP}}\delta_{\mathrm{CGP}} - \delta_{\mathrm{CGP}}^2} \tag{74}$$

It is difficult to derive a precise relationship between δ_{rp} and δ_{CGP}. However, since the particle–wall contact conduction is small compared to particle–fluid–wall, a very simple assumption is used here. We use the maximum overlap instead of the otherwise variable overlap during the particle–wall collision. Since this collision is modeled here by a mass-spring-dashpot system, as derived in Buist et al. (2016), the maximum overlap can be expressed as

$$\delta_{\max} = \sqrt{\frac{v_0^2 m_{\text{eff}}}{k_{\text{n}}}} f(e) \tag{75}$$

where v_0 is the initial collision velocity, k_{n} is the spring constant, and m_{eff} is effective mass. The expression of $f(e)$ can be evaluated by

$$f(e) = e^{\frac{-\zeta}{\sqrt{1-\zeta^2}}\cos^{-1}(\zeta)}, \quad \zeta = \frac{-\ln(e)}{\sqrt{\pi^2 + \ln(e)^2}} \tag{76}$$

Since the same spring constant k_{n} used in DEM is also used in our CGPM simulation, the overlap is calculated as

$$\delta_{\text{CGP}} = \delta_{\text{rp}} \sqrt{k^3} \frac{f(e_{\text{CGP}})}{f(e_{\text{rp}})} \tag{77}$$

The contact radius of the CGP is

$$r_{\text{c}}^{\text{CGP}} = k^{5/4} \left(\frac{f(e_{\text{CGP}})}{f(e_{\text{rp}})} \right)^{1/2} \sqrt{2 r_{\text{rp}} \delta_{\text{rp}} - k^{1/2} \frac{f(e_{\text{CGP}})}{f(e_{\text{rp}})} \delta_{\text{rp}}^2} \tag{78}$$

In a DEM simulation, the particle–wall overlap is usually much smaller than the collision radius. The contact radius of the coarse-grained system can, therefore, be approximated as

$$r_{\text{c}}^{\text{CGP}} = k^{5/4} \left(\frac{f(e_{\text{CGP}})}{f(e_{\text{rp}})} \right)^{1/2} r_{\text{c}}^{\text{rp}} \tag{79}$$

The error introduced here due to coarsening depends on the restitution coefficient of the original system and scale ratio, k. We must also consider the particle–wall collision frequency. Thus, the equation for particle–wall contact conduction is

$$Q_{i,\text{w}}^{\text{pwc}} = \frac{Q_{\text{CGP}}^{\text{pwc}}}{k^{9/4}} \left(\frac{f(e_{\text{rp}})}{f(e_{\text{CGP}})} \right)^{1/2} \tag{80}$$

Similar to particle–wall heat conduction, the particle–particle conduction involves particle–fluid–particle conduction and particle–particle contact conduction. The relationship between the coarse-grained system and the original system can be derived by following the same procedure as for particle–wall conduction. Considering a CGP representing k^3 number of real particles, only k^2 real particles on the surface can transfer heat to other CGP either through particle–fluid–particle conduction or contact conduction. Thus, the relationship derived for particle–wall conduction is still applicable to particle–particle conduction

$$Q_{\text{rp}}^{\text{pfp}} = Q_{\text{rp}}^{\text{pfp}}/k^2 \tag{81}$$

$$Q_{\text{rp}}^{\text{pp}} = \frac{Q_{\text{CGP}}^{\text{pp}}}{k^{9/4}} \left(\frac{f(e_{\text{rp}})}{f(e_{\text{CGP}})} \right)^{1/2} \tag{82}$$

In a DEM simulation, r_c is usually overestimated due to the adoption of a smaller Yong's module to achieve large solids time step. Thus, it is difficult to calculate this term precisely even in DEM. Zhou et al. (2010) proposed a correction coefficient, c, to reduce this overestimation for DEM simulation with a Hertzian model. This correction has been extended to the linear spring-dashpot model in MFIX-DEM as

$$r_{\text{c},0} = \left(\frac{k_{\text{n}} r_{\text{c}}}{k_{\text{n},0}} \right)^{2/3} \tag{83}$$

where $r_{\text{c},0}$ is the reduced contact radius. k_{n} is the spring constant used in the simulation, $k_{\text{n},0}$ is the spring constant derived from Yong's modules of particles through

$$\begin{aligned} k_{\text{n},0} &= \frac{4}{3}\sqrt{r_{\text{eff}}}E_{\text{eff}} \\ \frac{1}{r_{\text{eff}}} &= \frac{1}{r_i} + \frac{1}{r_j} \\ \frac{1}{E_{\text{eff}}} &= \frac{1-v_i^2}{E_i} + \frac{1-v_j^2}{E_j} \end{aligned} \tag{84}$$

where r_i and r_j are the particle radii of particles i and j. E_i and E_j are Yong's modules of particles i and j. v_i and v_j are Poisson ratios of particles i and j.

2.3 Extension to Time-Driven Hard-Sphere Contact Model

As will be shown later in this chapter, the main reason to investigate the hard-sphere collision model is the significant computer time savings it can achieve over the traditional soft-sphere model in DEM by allowing a larger solids time step yielding a speed-up of 20–60 times. In TDHS, CFD is used to solve for the fluid flow on a fixed Eulerian mesh as mentioned earlier, while the motion of particles is solved by a Lagrange tracking method. The difference between TDHS and DEM is that the particle–particle interactions are modeled based on momentum conservation laws rather than a linear spring-dashpot. As proposed by Ouyang and Li (1999), the movement of particles in gas–solids flows is decomposed into suspension and collision processes. The suspension process accounts for external forces like gravity and interphase forces, while the collision process handles the instantaneous binary collisions.

For suspension processes, the velocities of particles are updated following Newton's law of motion

$$m_{\mathrm{p}}\frac{d\mathbf{v}_{\mathrm{p}}}{dt}=m_{\mathrm{p}}\mathbf{g}-\frac{\pi}{6}d_{\mathrm{p}}^{3}\nabla P_{\mathrm{f}}+\frac{\beta\left(\mathbf{v}_{\mathrm{f}}-\mathbf{v}_{\mathrm{p}}\right)}{1-\varepsilon_{\mathrm{f}}}\frac{\pi}{6}d_{\mathrm{p}}^{3} \tag{85}$$

where m_{p} is the mass of the particle, $\mathbf{v}_{\mathrm{p}}$ its velocity, and d_{p} its diameter. On the right-hand side, the forces considered include gravity, pressure gradient, and drag.

For collision processes, the velocities of particles are updated by instantaneous binary collisions. When particle i and j are in collision, the variation of momentum during impact is

$$\Delta\mathbf{P}=-m_{\mathrm{e}}(1+e_{\mathrm{n}})\mathbf{v}_{\mathrm{n}}-\frac{2}{7}m_{\mathrm{e}}(1+e_{\mathrm{t}})\mathbf{v}_{\mathrm{t}} \tag{86}$$

where m_{e} is the effective mass of two colliding particles and can be calculated as

$$m_{\mathrm{e}}=\frac{m_i m_j}{m_i+m_j} \tag{87}$$

$\mathbf{v}_{\mathrm{n}}$ and $\mathbf{v}_{\mathrm{t}}$ are relative velocity in normal and tangential directions and can be calculated as

$$\mathbf{v}_{\mathrm{n}}=\left|\mathbf{v}_{\mathrm{c}}\right|\mathbf{n} \tag{88}$$

$$\mathbf{v}_{\mathrm{t}}=\mathbf{v}_{\mathrm{c}}-\mathbf{v}_{\mathrm{n}} \tag{89}$$

$$\mathbf{v}_{\mathrm{c}}=\mathbf{v}_j-\mathbf{v}_i-\left(r_i\boldsymbol{\omega}_i+r_j\boldsymbol{\omega}_j\right)\times\mathbf{n} \tag{90}$$

$$\mathbf{n}=\frac{\mathbf{x}_i-\mathbf{x}_j}{\left|\mathbf{x}_i-\mathbf{x}_j\right|} \tag{91}$$

where $\mathbf{x}_i$ and $\mathbf{x}_j$ are positions of particle i and j. To consider the friction limit on tangential momentum variation, Eq. (86) is rewritten as

$$\Delta\mathbf{P} = -m_e(1+e_n)\mathbf{v}_n + \mathbf{J}_t \tag{92}$$

$$\mathbf{J}_t = -\max\left(\frac{2}{7}m_e(1+e_t)|\mathbf{v}_t|, \mu m_e(1+e_n)|\mathbf{v}_n|\right)\frac{\mathbf{v}_t}{|\mathbf{v}_t|} \tag{93}$$

where μ is sliding friction coefficient; then, the velocity of particle i and j can be updated as

$$\mathbf{v}_i = \mathbf{v}_i - \frac{\Delta\mathbf{P}}{m_i} \tag{94}$$

$$\mathbf{v}_j = \mathbf{v}_j + \frac{\Delta\mathbf{P}}{m_j} \tag{95}$$

As for particle rotation, it can be calculated as

$$\boldsymbol{\omega}_i = \boldsymbol{\omega}_i - \frac{r_i\mathbf{n}\times\mathbf{J}_t}{I_i} \tag{96}$$

$$\boldsymbol{\omega}_j = \boldsymbol{\omega}_j - \frac{r_j\mathbf{n}\times\mathbf{J}_t}{I_j} \tag{97}$$

where I is the moment of inertia and is equal to $m_p d_p^2/10$.

The grid-based neighbor list method is used for the collision detection process (Garg et al., 2012a). For each particle i, an array stores its neighbor particles. During the calculation, the program loops over all the neighbors of particle i. If they overlap while moving closer, the collision is identified and is handled by the previously described procedure. Since the collision is handled in a predefined sequence, the algorithm cannot ensure all the colliding particles separate from each other after each collision. Thus, there could be some unphysical large overlaps among particles in the dense phase where collision frequency is high. To reduce large overlaps created during the hard-sphere calculation step, a novel velocity correction procedure is carried out right after the collision process. It loops over all the particles still in collision after the traditional hard sphere collision process. This correction algorithm corrects particle velocity in such a way that the largest overlaps are reduced the most by emphasizing particles with the largest overlaps and largest colliding velocity. To achieve this reduction in overlap, this model requires two unknown parameters that can be adjusted to achieve a reasonable overall overlap. We found that these parameters can be expressed by analogy with a spring-dashpot model. This provided reasonable overlaps that were comparable to those achieved by DEM simulation. The whole algorithm is shown in Fig. 6.

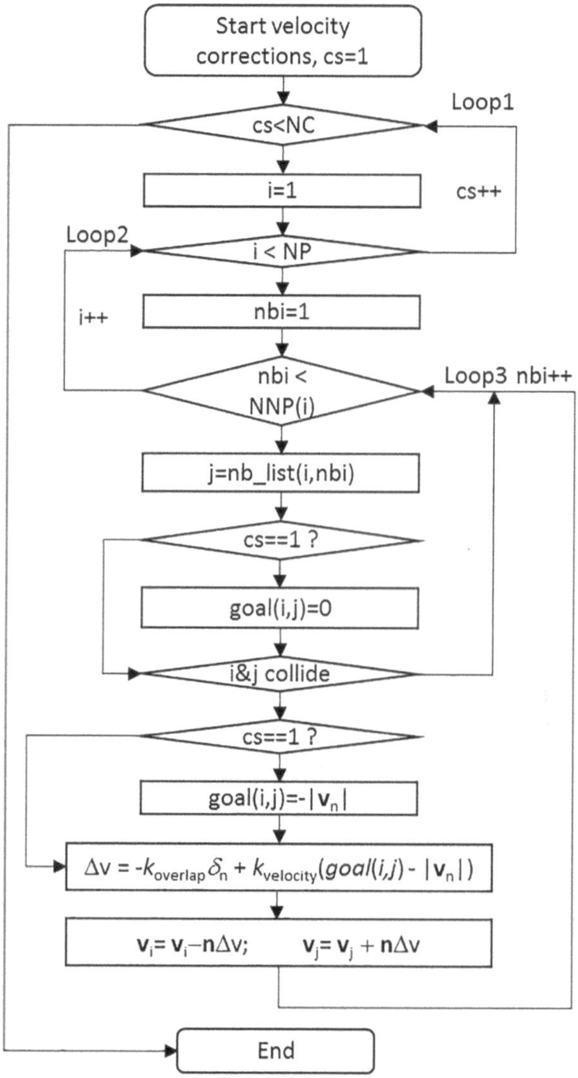

Fig. 6 Velocity correction procedure conducted right after the collision process.

As shown in Fig. 6, Loop 1 indicates that the correction loops over user-defined maximum correction steps, NC. Loop 2 and Loop 3 indicate that for each particle, the algorithm will loop over all its neighboring particles. If a collision is detected, which means two particles are overlapping and moving toward each other, a corrected velocity is calculated. The calculated correction term is a penalty function of overlap, δ_n, and the desired relative velocity. It should be noted that in dense flows where many overlaps exist between particles, this algorithm is only designed to reduce the largest overlaps that will otherwise be present in the next time step. For these dense flows, the exit out of this iteration loop is usually obtained by reaching the maximum correction steps, which typically comprise a small value of 2–3 steps. The two stiffness parameters, k_{overlap} and k_{velocity}, can be selected with an analogy to soft-sphere DEM by setting $k_{\text{overlap}} = \Delta t k_n / m_p$ and $k_{\text{velocity}} = \Delta t \eta_n / m_p$. Where k_n is the normal spring constant and η_n is the damping coefficient.

Although the CFD–TDHS method is much faster than traditional CFD–DEM by using a larger time step, it is still not fast enough for industrial- or pilot-scale problems because of the large number of particles in these systems. However, CFD–TDHS can be easily combined with the parcel-based method. Also, since the interaction between particles is modeled by momentum conservation laws, there is no need to derive the scaling laws for spring stiffness as used in DEM. The previous scaling laws for the restitution coefficient are based on the fact that the particle collision frequency is significantly reduced in the coarse-grained method; thus, these laws are still valid for the hard sphere method. To distinguish it from particle-based method, the coarse-grained hard sphere method is abbreviated as CGHS. The CGHS is validated through the simulation of a pilot-scale CFB riser, which will be discussed in Section 3.5.

3. VERIFICATION AND VALIDATION STUDIES

The methods discussed in Section 2 are all implemented in the freely available open-source multiphase flow simulation code MFIX. The MFIX-DEM code has been involved in some verification studies (Garg et al., 2012b) as well as validations (Li et al., 2012b) and has been widely used by the multiphase flow modelling and simulation community. In this section, coarse-grained CFD–DEM, TDHS, CFD–TDHS, and CFD–CGHS are subjected to some validation studies either by comparing their predictions with available experimental data or with more precise simulation methods like CFD–DEM or DNS. The first step is verifying the ability of the TDHS method to simulate dense granular flow and particle packing. The second step also involves the V&V of CFD–TDHS by simulating a small bubbling fluidized bed and comparing the results with CFD–DEM, DNS, and experimental data. The third step is verifying the major assumption of coarse graining CFD–DEM and CFD–CGHS by simulating a small bubbling fluidized bed and comparing the results with CFD–DEM. The fourth step is the V&V of the heat transfer models presented in this study for coarse-grained CFD–DEM by simulating a small bubbling fluidized bed and comparing the results with CFD–DEM and experimental data. Finally, the CFD–CGHS is validated by simulating a pilot-scale CFB riser where experimental data are available.

3.1 Verification Study of the Time-Driven Hard Sphere Model for Dense Granular Flow

The TDHS was implemented in the MFIX CFD code and verified for a simple homogeneous cooling system, particle settling case, and bubbling fluidized bed in a recent publication (Lu et al., 2017d). As explained in Section 2.3, the TDHS is unable to handle multibody collisions and will result in unphysical overlaps for dense granular flow. In this verification study, a simple particle settling system is simulated with DEM and TDHS. As shown in Fig. 7, initially, the particles are randomly distributed in the simulation box with a small random velocity or granular temperature. Then, under the pull of Earth's gravity, the particles settle to the bottom wall of the simulation domain. The detailed simulation parameters are listed in Table 1.

Fig. 8 compares the results of DEM, original TDHS method, and TDHS with our proposed corrections as detailed in Section 2.3. We first notice that the initial random velocity of the particles, as seen in Fig. 7, is quickly and significantly reduced due to the energy-dissipating collisions. Qualitatively, this figure shows that the original TDHS cannot predict a correct packing

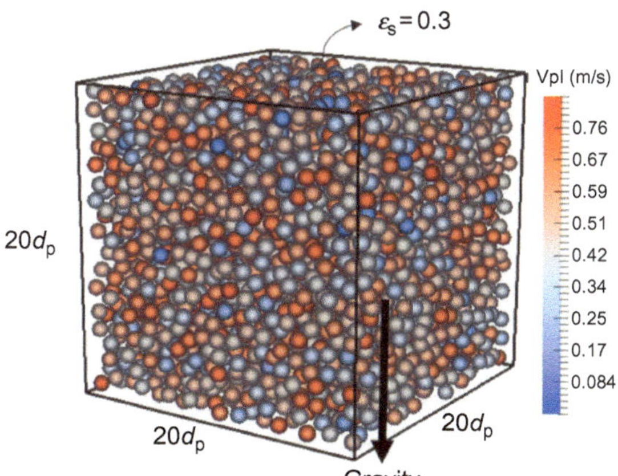

Fig. 7 Initial conditions and system geometry of particles settling under gravity.

Table 1 Simulation Parameters of Particles Settling System

Parameter or Property	**Values**
Particle diameter, d_p (m)	1.0×10^{-3}
Particle density, ρ_p (kg/m^3)	2500
Particle spring constant, k_n (N/m)	1.0×10^2
Particle restitution coefficient, e_n	0.95
Particle friction coefficient, μ	0.10
Particle restitution coefficient, e_t	0.95
Particle wall friction coefficient, μ_w	0.10
Average solid volume fraction, ε_s	0.30
Domain size, $W \times H \times D$	$20d_p \times 20d_p \times 20d_p$
DEM time step (s)	1.0×10^{-5}
TDHS time step (s)	1.0×10^{-4}, 1.0×10^{-5}
TDHS velocity correction steps	0,3,11

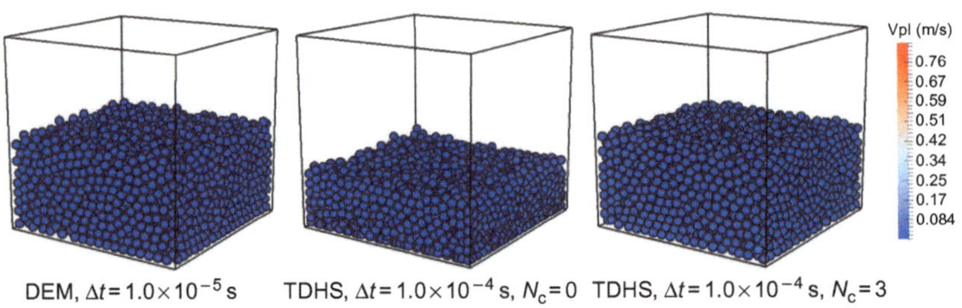

Fig. 8 Particle positions after 500 ms, N_c indicates the number of velocity corrections.

density without our proposed velocity correction term. This is because, for this settling case, the dominate contact mechanism is multibody collision, which is beyond the capability of the original TDHS. Without our proposed corrections, the particles will keep collapsing, which will cause many particles to completely overlap. However, this unphysical overlap can be reduced by the proposed velocity correction algorithm. This correction term prioritizes the separation of colliding particles with largest overlaps and largest collision velocities. This algorithm may result in small overlaps and/or a minor increase of overlaps for other particles, which was insignificant for the macroscopic results. A simple demonstration of this algorithm was conducted for a 1D simulation of three particles under gravity (Lu et al., 2017d).

We should stress the fact that predicting the wrong solids packing has a profound impact on the fluidization behavior of powders. Simply put, predicting the wrong solids packing implies the wrong value of computed drag force, which implies the wrong fluidization behavior. Therefore, reducing errors of void fraction calculation is essential for a method such as TDHS to have any practical application in fluidization.

Apart from the above qualitative analysis, we also quantitatively analyzed the influence of the number of correction steps and the computation time steps on the settling process by comparing the normalized weight center, which relates directly to the solids packing fraction. As shown in Fig. 9, for DEM simulations, the normalized weight center stabilizes around a value of 0.5 after a settling time of only about 0.06 s (recall that this simulation is conducted in a small system geometry). Considering that the initial solids volume fraction is 0.3, the final packing solids volume fraction is about 0.6, which is a reasonable value for random packing of spheres. Before this maximum packing value is reached, or before 0.06 s, the system is dominated by binary collisions and, even without any correction, the original TDHS can predict the same profile as DEM. However, after 0.06 s, the system is densely packed and dominated by multibody enduring contacts. The predicted profile obtained with the original TDHS keeps

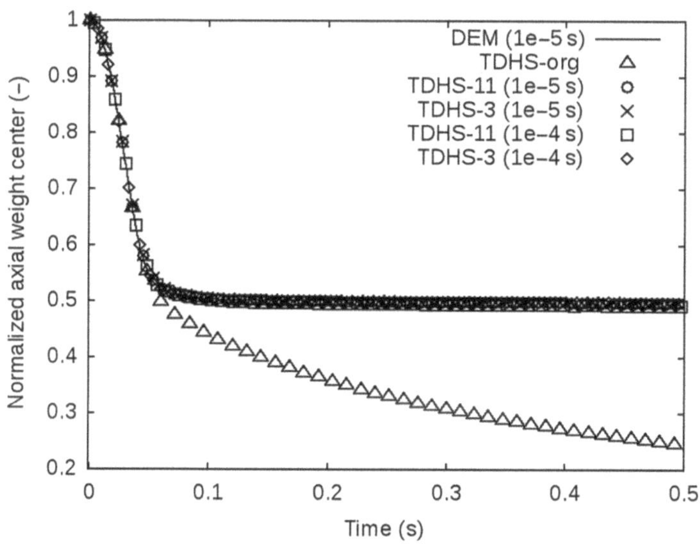

Fig. 9 Particle normalized weight center with different number of correction and time steps. *Modified from Lu L, Li T, Benyahia S: An efficient and reliable predictive method for fluidized bed simulation.* AIChE J *63, 5320–5334, 2017d, https://doi.org/10.1002/aic.15832 with permission for the reprinted content. Copyright (2017) Wiley.*

decreasing and at 0.5 s the predicted normalized weight center is about 0.25 that of the initial value. This means the packing density is about 1.2, which is physically impossible. If our proposed velocity correction is used, then the TDHS can avoid this unphysical phenomenon and predicts a similar profile and packing value as DEM while using an order of magnitude larger computation time step. Also, the influence of the number of correction steps and computation time steps on the final packing density is small. This means that the proposed modification to the TDHS method can be used with a much larger time step than DEM, which will allow for a large increase in the simulation speed.

To further analyze the influence of the number of correction steps and computation time steps, we also compare the normalized kinetic energy of the system predicted by different methods. As shown in Fig. 10, the kinetic energy predicted by the DEM simulation can be reduced to about 10^{-9} of the initial value in the system. Theoretically this value should tend to zero, and the small value obtained by DEM is mainly caused by numerical errors. For TDHS, the computation time step has a major influence on the results. For larger time steps (1.0×10^{-4} s), the kinetic energy is only dissipated to about 0.001 of the initial system value. This is because the TDHS algorithm does not resolve the particle collision process as in DEM. During the correction process, the potential energy stored in overlaps is transformed to kinetic energy. Since a larger time step is used, numerical errors occur during

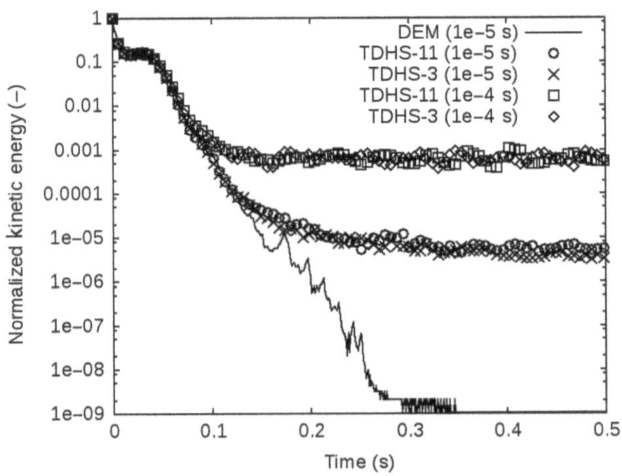

Fig. 10 Particle normalized kinetic energy with different numbers of correction and time steps. *Modified from Lu L, Li T, Benyahia S: An efficient and reliable predictive method for fluidized bed simulation.* AIChE J *63, 5320–5334, 2017d, https://doi.org/10.1002/aic.15832 with permission for the reprinted content. Copyright (2017) Wiley.*

this transformation that generate an excess of kinetic energy. When this numerically generated energy reaches a value comparable to that dissipated by the collision process, the total kinetic energy fluctuates around a stable value. This numerical error can be significantly reduced by reducing the solids time step. As an example, decreasing the time step by an order of magnitude to a value of 1.0×10^{-5} s can reduce the energy dissipation to about 10^{-5} of the initial system. This reduction in numerical errors with the smaller time step of the corrected TDHS method can be useful in engineering applications. For example, a larger time step may be used at the first stages of a simulation and then reduced to improve simulation results if more computational resources become available or more accurate simulation results are required. Nevertheless, it is obvious to us that the gain in speed obtained with a larger solids time step compensates for the small loss of accuracy. Another observation is that the number of correction steps exerts only a minor influence on the final results. This is because most of the large overlaps are reduced after a few corrections, and overlaps still existing after the corrections in this time step will be further reduced in the next step.

Now we may conclude that using a velocity correction algorithm significantly reduces the unphysical overlaps and that TDHS can predict results similar to results with the DEM method for dense granular flow. Thus, this method can be used to simulate dense and CFBs where multibody collisions may also play an important role in stagnation zones and dense clusters.

3.2 Validation of Time-Driven Hard Sphere Model for Bubbling Fluidized Bed

Although CFD–TDHS has been used to simulate gas–solids flows (Helland et al., 2005; Ouyang and Li, 1999), this method lacks thorough V&V by comparing its predictions with experimental measurements and other more precise simulation methods like DEM and DNS. In this section, the V&V of CFD–TDHS is conducted through the simulation of a small fluidized bed. The time-averaged solids flux spatial distributions is compared with experimental data as well as with DNS results using the immersed boundary method (IBM) (Tang et al., 2016). The simulation domain and boundary conditions are shown in Fig. 11. Initially, the particles are randomly generated in the whole physical domain. A settling period follows, to allow the particles to pack under the force of gravity at the bottom of the bed, then the fluidized bed simulation starts. During this simulation, air flows uniformly into the reactor at the bottom of the bed with a superficial velocity of 2.4 m/s, while the outlet located at the top of the bed is set at a constant atmospheric pressure of 101 kPa. The system geometry consists also of left,

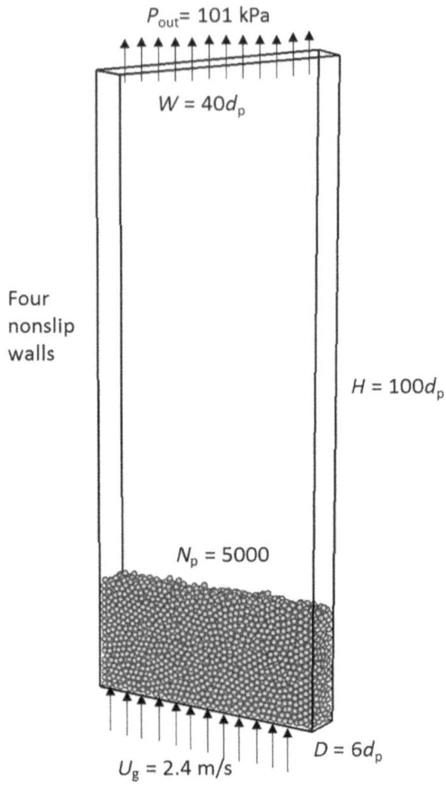

Fig. 11 Simulation setting of small bubbling fluidized bed.

right, front, and back walls, which are all set to nonslip walls for the gas phase. The Syamlal and O'Brien drag model is used. We stress that the same initial and boundary conditions are used for both CFD–DEM and CFD–TDHS simulations that were conducted in this study. Other simulation parameters are listed in Table 2. Further details of this case can be found in a recently published paper (Lu et al., 2017d).

The instantaneous particle positions colored with their velocities are presented in Fig. 12. The predicted flow patterns obtained with CFD–DEM and CFD–TDHS are reasonably similar. Both are characterized by a large upward-moving bubble in the center and downward flow of particles near the side walls. This fluidization behavior of Geldart D particles in a shallow bed may exhibit slugging for deeper beds, which is typically observed in the fluidization of such coarse particles. For brevity, only one instant screenshot is shown here, while the fluidized bed is simulated for 40 s with 800 saved frames, and the corresponding Video 2 in the online version at

Table 2 Simulation Parameters in Bubbling Fluidized Bed

Parameter or Property	**Value**
Particle diameter, d_p (m)	2.5×10^{-3}
Particle density, ρ_p (kg/m^3)	2526
Particle spring constant, k_n (N/m)	8.0×10^2
Particle restitution coefficient, e_n	0.97
Particle friction coefficient, μ	0.1
Particle restitution coefficient, e_t	0.33
Particle wall friction coefficient, μ_w	0.2
Superficial gas velocity, u_g (m/s)	2.4
Domain size, $W \times H \times D$	$40d_p \times 100d_p \times 6d_p$
CFD grid number, $N_W \times N_H \times N_D$	$20 \times 50 \times 3$
Solid time step (s)	1.0×10^{-5} (DEM), 1.0×10^{-4} (TDHS)
CFD time step (s)	1.0×10^{-4}
Particle number N_p	5000

https://doi.org/10.1016/bs.ache.2017.12.001 is uploaded as supplementary material. Of course, these two methods do not predict the exact same results because particle–particle interactions are handled differently, and a much larger time step is used for CFD–TDHS.

An interesting quantity to measure is the Favre-averaged velocity, or the solids volume flux, which is the time-averaged value of the product of the instantaneous solids velocity and volume fraction. It is compared quantitatively with experimental and DNS data in Fig. 13. The simulation results are averaged from 10 to 40 s with a frequency of 20 Hz, while the experimental data and IBM results are reproduced from the published literature (Tang et al., 2016). At a height of 0.025 m above the gas distributor, the simulated results compare well with both experimental data and DNS simulation results. Both CFD–DEM and CFD–TDHS predict quantitatively the correct trends with an upward flow in the center of the bed and a downward flow near the wall region. At a height of 0.055 m, the simulation still predicts the correct quantitative trends of an upward flow of solids in the center of the bed with an underpredicted downflow of solids near the wall region compared to experimental measurements. This may have been caused by

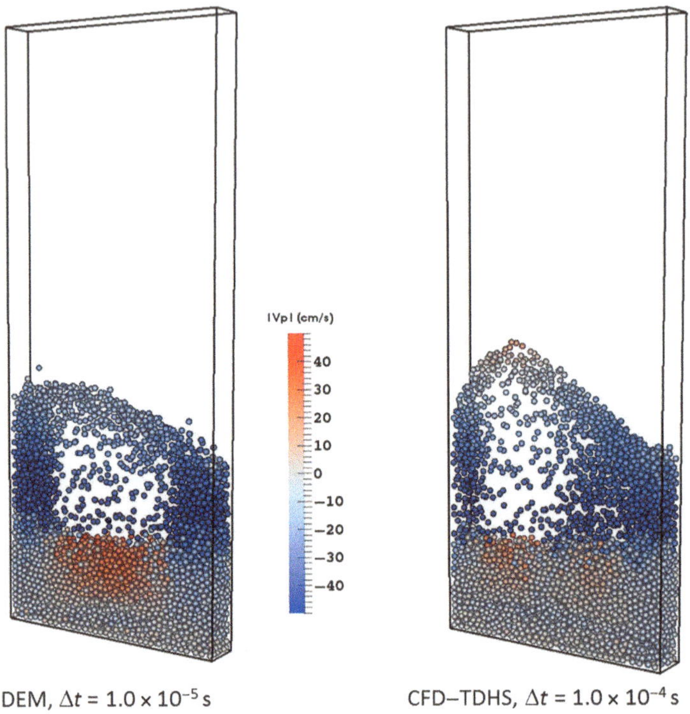

Fig. 12 Instantaneous particles distribution obtained with CFD–DEM and CFD–TDHS.

inaccurate modelling of particle–wall interactions and near-wall fluid flow field. Nevertheless, the CFD–TDHS predictions are very similar to those obtained with CFD–DEM, which evidently demonstrates that CFD–TDHS is at least as accurate as CFD–DEM. At a height of 0.085 m, all the numerical methods failed to predict the correct quantitative experimentally measured profile of solids Favre velocity. Like the results obtained at a height of 0.055 m, the simulation results at 0.085 m underpredict the negative solids flux experimentally observed near the wall as well as the upward solids flow at the center of the bed. However, the result of CFD–TDHS is still very similar to that obtained with our CFD–DEM, which is also in reasonable agreement with the published DNS data. We should mention here that all the numerical techniques compared here are CFD-based, but we recognize that DNS data are the most accurate because it makes no assumption on any fluid–particle interaction terms as the flow around each particle is fully resolved. All these terms, such as fluid–particle drag force, added mass, and lift forces (not included in this study), are explicitly calculated in DNS without any assumption. Obviously the best accuracy we can hope for our simulation results to achieve is that of DNS. Achieving similar quantitative trends as those

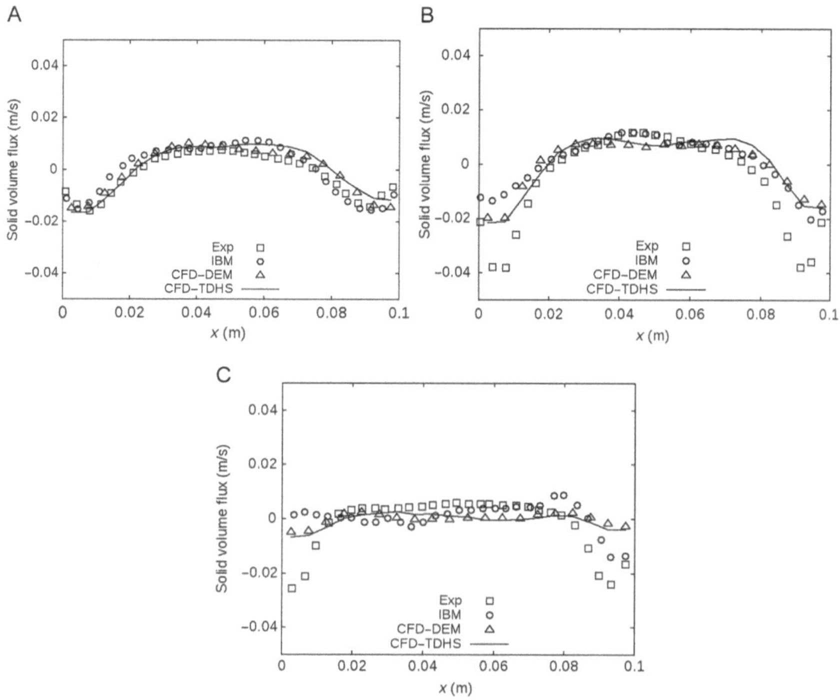

Fig. 13 Solids volumetric flux profile at different heights simulated with CFD–DEM and CFD–TDHS. (A) Height = 0.025 m, (B) height = 0.055 m, (C) height = 0.085 m. *Modified from Lu L, Li T, Benyahia S: An efficient and reliable predictive method for fluidized bed simulation.* AIChE J *63, 5320–5334, 2017d, https://doi.org/10.1002/aic.15832 with permission for the reprinted content. Copyright (2017) Wiley.*

obtained with DNS is, therefore, satisfactory even though some quantitative disagreement with experimental measurements persists.

In addition to first-order statistics, we also compared the particles fluctuating energy, also known as granular temperature, which is a second-order statistical quantity of interest. It can be estimated as

$$\Theta = \frac{1}{3}\left(\Theta_x + \Theta_y + \Theta_z\right) \tag{98}$$

$$\Theta_x = \frac{1}{N}\sum_{i=1}^{N}\left(v_x - \overline{v}_x\right)^2 \tag{99}$$

$$\Theta_y = \frac{1}{N}\sum_{i=1}^{N}\left(v_y - \overline{v}_y\right)^2 \tag{100}$$

$$\Theta_z = \frac{1}{N}\sum_{i=1}^{N}\left(v_z - \overline{v}_z\right)^2 \tag{101}$$

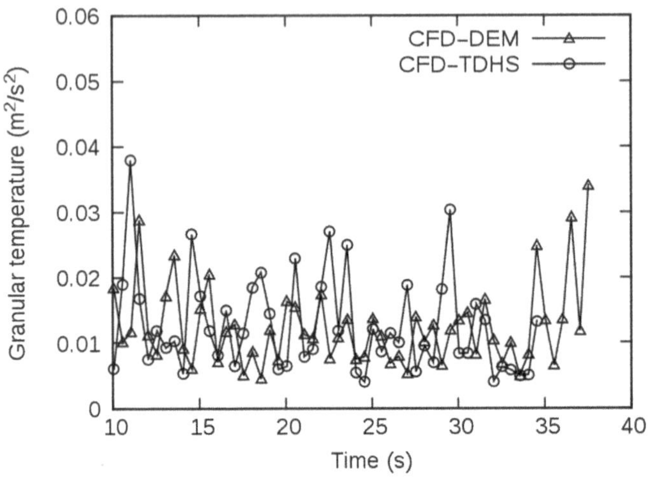

Fig. 14 Granular temperature profiles predicted with CFD–DEM and CFD–TDHS methods. *Modified from Lu L, Li T, Benyahia S: An efficient and reliable predictive method for fluidized bed simulation.* AIChE J *63, 5320–5334, 2017d, https://doi.org/10.1002/aic.15832 with permission for the reprinted content. Copyright (2017) Wiley.*

To calculate this second-order statistical quantity, first the velocities of all particles in the simulation domain were spatially averaged. Then, the square of the velocity difference between each individual particle and the domain-averaged value was again spatially averaged and plotted against time. As shown in Fig. 14, both the transient behavior of the particle fluctuating energy as well as the time-averaged values were very similar for DEM and the newly developed CFD–TDHS method. Note that this higher statistical quantity of interest was not measured in the published literature for this case; no direct comparison between our current results and experimental/DNS data is possible. Taking the results of DEM as a reference value, the relative error for CFD–TDHS was 6.4%. These quantitative comparisons of DEM and TDHS support the conclusion that these two techniques yield similar low and high-order statistical predictions. This strongly supports the TDHS technique because it gives similar accuracy as DEM but with a much faster (20–60 times faster) execution speed.

3.3 Verification Study of Coarse-Grained Soft-Sphere and Hard-Sphere Contact Models

Previous results indicate that the CFD–TDHS predictions are similar to those obtained using CFD–DEM with the benefit of a much larger time step.

However, for large-scale systems, the computation speed is still not fast enough to obtain predictions in a reasonable time frame due to the large number of particles in these systems. To further reduce the computation cost in CFD–DEM and CFD–TDHS, several real particles are lumped into a computational parcel, which tracks fewer parcels and their collisions. In this section, a virtual bubbling fluidized bed is simulated with CFD–DEM, coarse-grained CFD–DEM, and CFD–CGHS methods. Since no experimental data are available for this hypothetical system, we consider this study as a verification exercise of the coarse-grained soft-sphere (DEM) and hard-sphere (TDHS) methods by comparing these coarse methods with the finer CFD–DEM approach that tracks each individual particle instead of computational parcels. The simulation domain is double the width, depth, and height of the small fluidized bed in Section 3.2, and so to keep the same particle volume fraction, the number of particles is increased by eight times (double in all three physical directions). Other parameters such as boundary conditions, gas and particle properties, and CFD grid size are the same as those used in Section 3.2. For simulations with coarse-grained CFD–DEM and CFD–CGHS, the parcel diameter is twice the real particle size, which indicates each computational parcel contains eight real particles.

The instantaneous particle distributions colored with their velocities obtained with three different methods are compared in Fig. 15. The qualitative

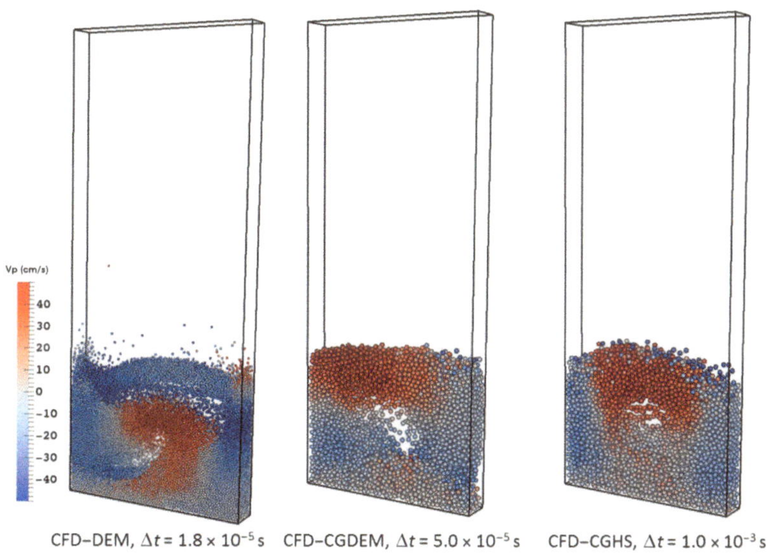

Fig. 15 Instantaneous particle distribution obtained with CFD–DEM, CFD–CGDEM, and CFD–CGHS.

predictions of gas and solids flow patterns obtained with coarse-grained CFD–DEM and CFD–CGHS compared well with the finer CFD–DEM simulation results. All the simulation results show a large bubble in the center of the pulsing fluidized bed. The dynamic flow patterns in this fluidized bed are recorded in Video 3 in the online version at https://doi.org/10.1016/bs.ache.2017.12.001 that is uploaded as a supplementary material available for the reader. Similar fluidization behavior can be observed in the animation for all three methods studied here. It is important to note that by lumping eight particles into a computational parcel, the number of computational parcels are reduced by a factor of eight, and that for the coarse-grained CFD–DEM the solids phase time step increases by a factor of about three due to the tracking of larger parcels. For CFD–CGHS, the solids phase time step is further increased by a factor of 20 due to the particle–particle collisions being handled by momentum conservation laws, and, thus, they do not need to be resolved like the soft-sphere collision method.

Macroscopic quantities of interest to engineers, such as solids velocity and solids volume fractions, are quantitatively compared in Fig. 16 for the three methods examined in this section. The simulation results are time-averaged from 20 to 50 s with a frequency of 20 Hz. At a height of 0.05 m above the gas distributor, both CFD–CGDEM and CFD–CGHS predict similar results to those obtained with finer CFD–DEM simulation. As typically observed in bubbling fluidized beds, particles tend to flow upward in the center of the bed where bubbles flow, and downward near the walls of the bed, which creates the large-scale recirculation of solids that is simply due to gas dragging particles against the constant pull of Earth's gravity. At a height of 0.11 m, the coarse methods still predict similar flow patterns as the finer method, with the

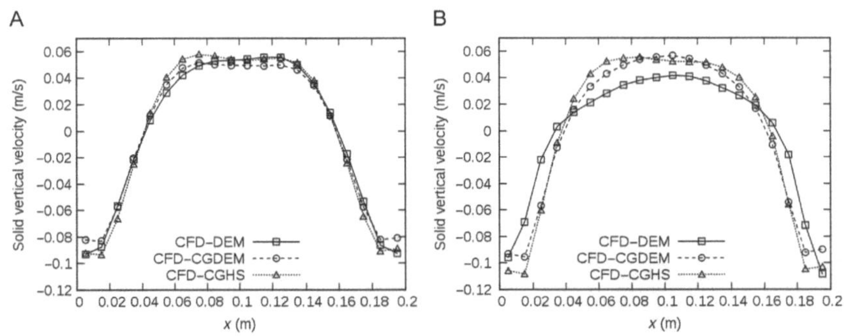

Fig. 16 Solids vertical velocity profile at different heights simulated with CFD–DEM, coarse-grained CFD–DEM, and CFD–CGHS. (A) Height = 0.05 m, (B) height = 0.11 m.

coarse-grained methods overpredicting the solids velocity both in the center upward flow and near the wall downward flow. This is because, in a coarse-grained system, the particles in a parcel are forced to move together, which reduces the degree of freedom in the system. The energy of the system also changes due to differences in the number of collisions. As with any other coarser approaches, significant gains in computation speed are always accompanied by some loss in accuracy. Carefully balancing speed with acceptable model-induced uncertainty is required by practicing engineers and researchers while using this or any other coarse method.

The solid volume fractions obtained with the three approaches used in this study are quantitatively compared in Fig. 17. At a lower height of 0.05 m, coarse-grained methods slightly underpredict the results obtained with finer CFD–DEM. However, at a height of 0.11 m a reversed phenomenon is observed where the coarse-grained methods now overpredict the results obtained with CFD–DEM. Overall, the larger axial gradients in solids volume fraction exist for the finer CFD–DEM method, which could occur because of a larger energy dissipation due to more frequent particle–particle collisions that increase inhomogeneity in the fluidized bed. Nevertheless, the quantitative differences between the fine and coarser methods are as small as other source of numerical errors, such as the time-averaging period.

As mentioned earlier, it is difficult to declare coarse methods fully verified and declare their use as validated to the practicing engineer or researcher. In general, it is fair to say that when speed is paramount, such as in many industrial applications, and some loss in accuracy can be tolerated, then the use of coarse techniques such as those explored here is fully warranted and must be pursued.

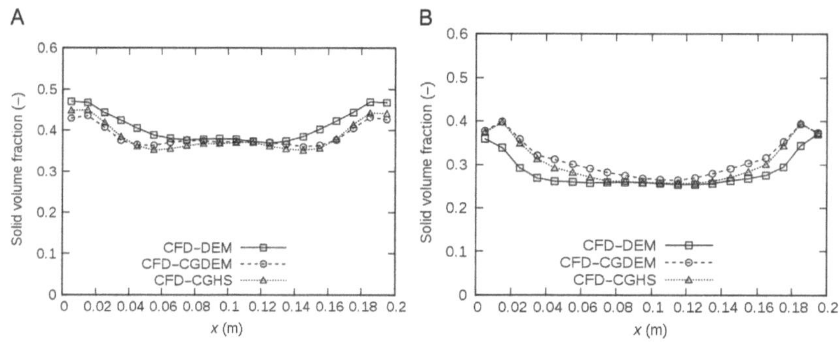

Fig. 17 Solids volume fractions at different heights simulated with CFD–DEM, coarse-grained CFD–DEM, and CFD–CGHS. (A) Height = 0.05 m, (B) height = 0.11 m.

3.4 Validation of Heat Transfer in Coarse-Grained Soft-Sphere Contact Model

To validate the heat transfer model developed in the methods section for coarse-grained CFD–DEM, a gas–solids fluidized bed (where experimental data are available) is simulated in this section. This fluidized bed has been experimentally investigated by Patil et al. (2015), and the physical domain is shown in Fig. 18. In this fluidized bed, glass beads initially at 90°C are fluidized by a cooler nitrogen gas at 20°C. The front plate of the bed is made of sapphire glass, while the back and side walls are made of aluminum. Experimentally, a circular nozzle with a diameter of 1.3 cm is located at the bottom of the bed. In neither the simulation nor the experimental run considered in this study, is gas injected through this nozzle, which is considered simply as a blocked area. In the simulation, however, this area is blocked with a rectangular 1.14 cm × 1 cm surface. There is an exception for simulation with a statistic weight of 27 (case W27), where the CFD grid is too large to resolve this small rectangular patch. Therefore, to ensure the same fluid flow rate through the bottom inlet, the superficial gas inlet velocity for case W27 is reduced from 1.33 to 1.20 m/s. The BVK drag model, described in the methods section, is used in this simulation. Other simulation

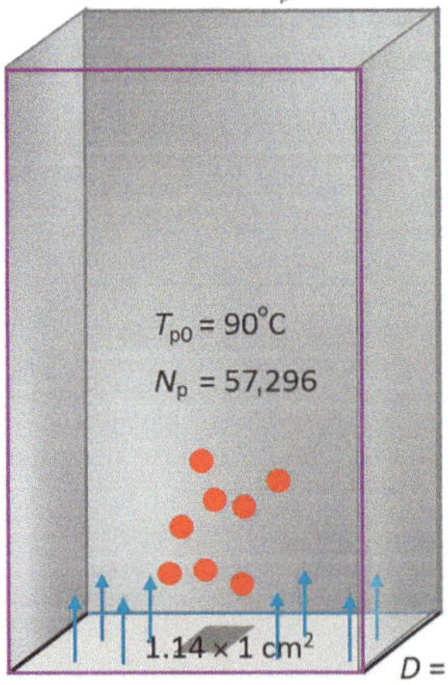

Fig. 18 Simulation setup for heat transfer bubbling fluidized bed. *Modified from Lu L, Morris A, Li T, Benyahia S: Extension of a coarse grained particle method to simulate heat transfer in fluidized beds.* Int J Heat Mass Transf *111:723–735, 2017c with permission for the reprinted content. Copyright (2017) Elsevier.*

parameters and physical properties are listed in Tables 3 and 4. More simulation details on this case can be found in a recently published paper (Lu et al., 2017c).

Since the aluminum constituting most of the side walls in this fluidized bed is of high thermal conductivity, a constant wall temperature boundary condition (293 K) is used for all the walls. The cooling line, representing

Table 3 Simulation Parameters and Physical Properties of Heat Transfer

Parameter or Property	Value
Particle diameter, d_p (m)	1.0×10^{-3}
Particle density, ρ_p (kg/m^3)	2500
Particle spring constant, k_n (N/m)	1.0×10^2
Particle restitution coefficient, e	0.97
Particle friction coefficient, μ	0.30
Particle heat capacity, C_p (J/kg K)	840
Particle thermal conductivity, k_p (W/m K)	1.40
Aluminum thermal conductivity, k_{Al} (W/m K)	238
Sapphire thermal conductivity, k_{Al} (W/m K)	40
Fluid heat capacity, C_f (J/kg K)	1010
Fluid viscosity, μ_f (kg/m s)	2.0×10^{-5}

Table 4 Simulation Parameters for CFD–DEM and CFD–CGDEM of Heat Transfer

	CFD-DEM	W8	W27
CFD grids, $N_x \times N_y \times N_z$	$35 \times 110 \times 6$	$35 \times 110 \times 6$	$16 \times 50 \times 3$
CFD time step, Δt (s)	1.0×10^{-4}	5.0×10^{-4}	5.0×10^{-4}
Number of CGP, N_{CGP}	57,296	7162	2122
Statistic weight, W	1	8	27
Solid time step, Δt (s)	1.27×10^{-5}	1.44×10^{-5}	2.64×10^{-5}
Gas layer diameter, r_{len} (m)	1.04×10^{-3}	2.08×10^{-3}	3.12×10^{-3}
Minimum conduction distance, s (m)	1.0×10^{-6}	2.0×10^{-6}	3.0×10^{-6}
Fluid inlet velocity, U_f (m/s)	1.33	1.33	1.20

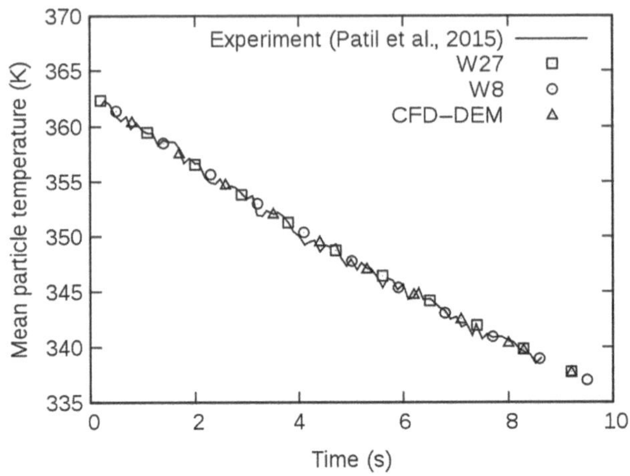

Fig. 19 Mean particle temperature profile simulated with different W and constant wall temperature boundary conditions. *Modified from Lu L, Morris A, Li T, Benyahia S: Extension of a coarse grained particle method to simulate heat transfer in fluidized beds. Int J Heat Mass Transf 111:723–735, 2017c with permission for the reprinted content. Copyright (2017) Elsevier.*

the mean temperature of particles, obtained experimentally, by DEM, and by coarse grained DEM with different statistic weight, is compared in Fig. 19. As shown, the mean particle temperature profile simulated with different W (recall that W stands for number of real particles in a parcel) compared well with experimental data. This indicates that the constant temperature boundary condition is close to reality. Also, the proposed heat transfer model for coarse-grained CFD–DEM can correctly predict the macroscale temperature transient variations for this particular case study.

To further verify the heat transfer model for coarse-grained CFD–DEM, different heat transfer mechanisms are quantitatively analyzed and compared with the finer method of CFD–DEM. In this fluidized bed, the effect of radiative heat transfer is ignored because the temperature is relatively low. The implementation and validation of heat transfer by radiation is left for future work and is not discussed in this chapter. The heat in particles is only lost by particle–fluid convection, particle–fluid–wall conduction, and particle–wall contact conduction. These different heat transfer mechanisms are quantitatively compared in Fig. 20 using numerical results obtained from CFD–DEM and coarse-grained CFD–DEM simulations. For the most important heat-transfer mechanism (particle–fluid convection term), the CFD–DEM results agree quantitatively well with those using the lower

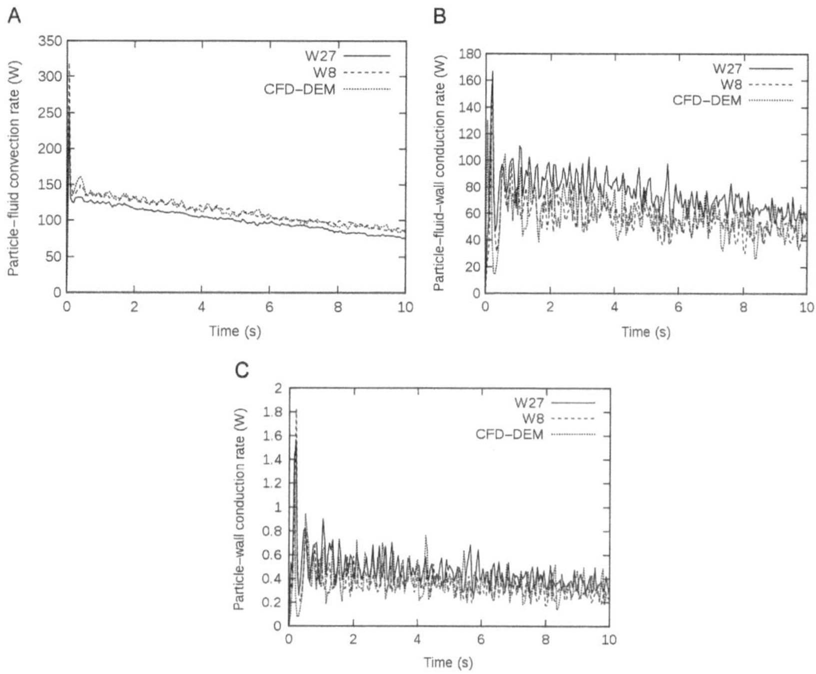

Fig. 20 Comparison of different heat transfer mechanisms calculated with different *W*. (A) particle–fluid convection, (B) particle–fluid–wall conduction, (C) particle–wall contact conduction. *Modified from Lu L, Morris A, Li T, Benyahia S: Extension of a coarse grained particle method to simulate heat transfer in fluidized beds. Int J Heat Mass Transf 111:723–735, 2017c with permission for the reprinted content. Copyright (2017) Elsevier.*

coarsening level (W8), while the results obtained with the most aggressive coarsening level (W27) underpredict the DEM results. Apart from errors inherent to coarse-graining, using coarser CFD grids, as in case W27, may have contributed further to this quantitative disagreement between coarse and finer methods. The rate of particle–fluid convection decreases gradually with time from an initial value of about 150–100 W after 10.0 s of simulation time, which was due to the temperature gradient between fluid and particles decreasing as the particle temperature decreases with time. The second most important heat transfer mechanism in this fluidized bed simulation is due to particle–fluid–wall conduction as indicated in Fig. 20B. For this mechanism, the results obtained with different grain-coarsening levels (*W*) are very similar. This proves the accuracy of our previously proposed particle–fluid–wall conduction model. Similar to the convective heat transfer term, the conductive heat transfer rate decreases from about 80 to 50 W indicating a decrease in the temperature gradient between particle and wall as time progresses. Finally, for particle–wall conduction, the heat transfer rate is the least important heat transfer mechanism with values less than 1 W. Although it seems that this value is negligible in this fluidized bed case, the results obtained with different statistical weights compared well with those obtained with DEM, indicating the accuracy of our proposed model for particle–wall conduction.

Although this case study and those previously conducted in this chapter are relatively simple because they deal mainly with small-scale fluidized beds, the case studies reveal that the coarse-grained method described in this chapter captures, with reasonable accuracy, both the hydrodynamic and heat transfer in fluidized beds. Next, we will be examining the validity of our proposed models in a pilot-scale fluidized bed where experimental data are available, and then we will explore several large-scale applications with interest to industry, including the heat transfer mechanisms studied here as well as mass transfer between particles and the fluid due to heterogeneous chemical reactions.

3.5 Validation of Coarse-Grained Hard-Sphere Model for Pilot-Scale CFB Riser

The new CGHS method developed in this chapter was used to simulate a pilot-scale CFB riser of 0.304 m diameter and 16.8 m height. This fluidized bed was designed and built at the National Energy Technology Laboratory (NETL) in Morgantown, WV, and details of the experimental setup can be found at https://mfix.netl.doe.gov/challenge-problem-iii-2010/. This CFB has a large-enough diameter and height to rival most available rigs published in the literature, and has been operating for a couple of decades with sophisticated measuring techniques capable of gathering most data of interest to engineers. For these reasons, its experimental data were selected for a challenge problem that was posed less than a decade ago to modelers around the world challenging them to make similar predictions with their models, thus probing the reliability and maturity of CFD models developed in academia, national labs, and industry. At that time, and without the development of the coarse-grained method, this kind of problem could only be investigated with TFM (Li et al., 2012a). Now, and with the development of CFD–CGHS, this pilot-scale problem can be simulated at the particle level with a computation speed comparable or even faster than TFM. The description of this CFB riser is presented in Fig. 21 showing that the main gas flows uniformly into the riser from the bottom distributor.

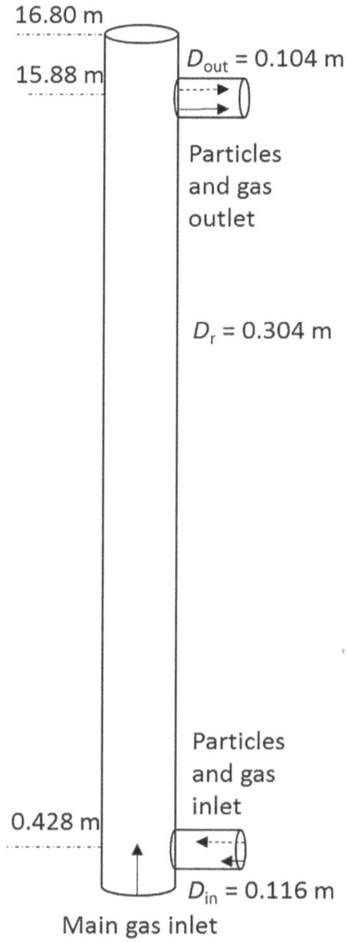

Fig. 21 Simulation setting of circulating fluidized bed riser.

The particles, along with some aeration gas, flows into the riser at the bottom side "L" valve pipe and exit freely from the "T" type outlet near the top of the riser. The Gidaspow drag model detailed in the methods section is used in this simulation. The detailed physical and simulation parameters are listed in Table 5. This case was also simulated in our previous research; other detailed analysis of results can be found in a recent publication (Lu et al., 2017d).

Fig. 22 shows the time-averaged void fraction (also called voidage) distributions in the riser as well as detailed particle positions near the solids main L-valve inlet. This figure shows a dense region near the walls of the bottom inlet and top outlet regions, as well as a less dense region in the middle

Table 5 Simulation Parameters of Circulating Fluidized Bed Riser

Parameter or Property	Value
Particle diameter, d_p (m)	8.02×10^{-4}
Particle density, ρ_p (kg/m^3)	863
Particle spring constant, k_n (N/m)	50
Parcel restitution coefficient, e_n	0.5, 0.14
Particle friction coefficient, μ	0.6
Particle restitution coefficient, e_t	0.3, 0.1
Particle wall friction coefficient, μ_w	0.6
Main inlet gas velocity, u_g (m/s)	6.81
Main inlet gas pressure (Pa)	124 339
Main inlet gas temperature (K)	294
Side inlet gas flow rate, Q_g (m^3/s)	2.37×10^{-2}
Side inlet gas pressure (Pa)	124 339
Side inlet gas temperature (K)	294
Side inlet solid mass flow rate, M_s (kg/s)	14.0
Outlet pressure (Pa)	105 000
CFD grid size, $d_x \times d_y \times d_z$ (m^3)	$0.016 \times 0.016 \times 0.016$
Solid time step (s)	5.0×10^{-5}, 1.0×10^{-4}
CFD time step (s)	5.0×10^{-4}, 1.0×10^{-3}

section of the riser. This "C" type distribution of solids concentration along the height of the riser is similar to that observed in the experiment. Near the bottom solids inlet, the incoming particles are moving toward the left side walls leading to an asymmetric distribution of particles. A similar asymmetric distribution can also be observed near the top region where the particles are moving toward the T-shaped outlet in the right side of the riser. For the middle sections, the distribution is more symmetric because the particles are less influenced by inlet and exit effects. The particles still exhibit a typical core-annulus flow behavior as indicated by a slightly denser region near the walls and more dilute region at the core of the riser. The instant particle distribution at time 40.0 s in the bottom region is also shown in this figure with colors referring to particle velocity. In this region, the incoming particles are

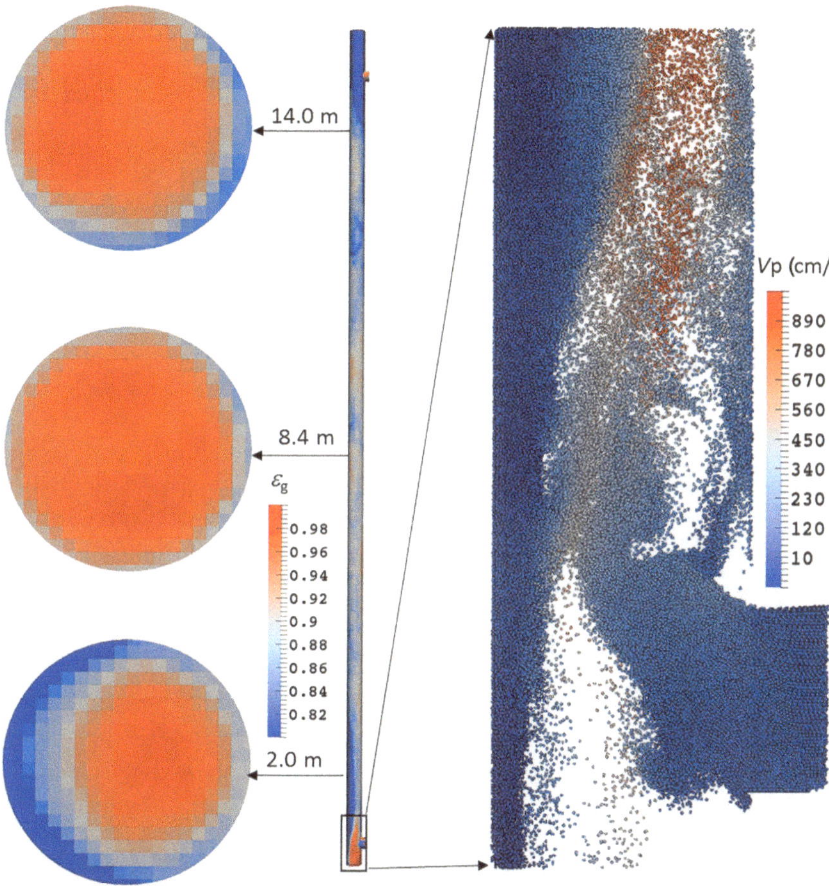

Fig. 22 Time-averaged voidage distribution and instant particle screenshot.

colliding with the downward particles near the right-side walls and are then dragged upward by the high inlet velocity from the bottom gas inlet distributor. Incoming particles, dense falling clusters of particles, and incoming air must battle their way into the riser in a very short and crowded area at the bottom section of the riser. This complex motion of particles can be observed in Video 4 and 5 in the online version at https://doi.org/10.1016/bs.ache.2017.12.001 included in the supplemental material.

The predicted radial profiles of time-averaged solids vertical velocity at different heights are compared with experimental measurements in Fig. 23. At a height of 6.23 and 8.88 m above the gas distributor, the fiber optical probe (FO) was used to measure the solids velocities and the high-speed particle image velocimetry (HSPIV) was used only at a height of 8.88 m to

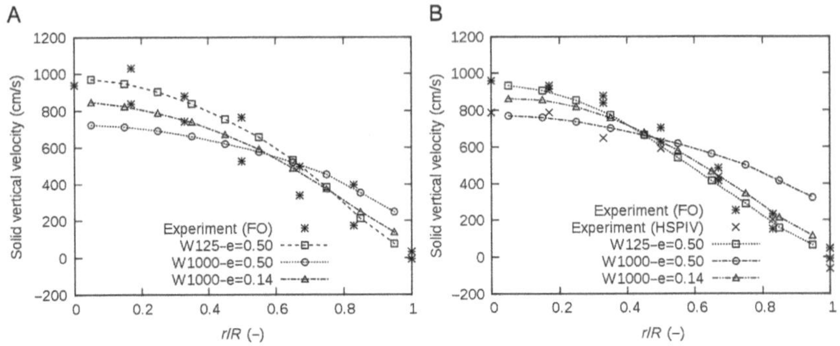

Fig. 23 Radial profiles of solid vertical velocity at different heights. (A) $H=6.23\,\text{m}$, (B) $H=8.88\,\text{m}$.

compare the different measuring techniques and to better assess the experimental measurement errors and uncertainties. Here the raw experiment data are plotted directly for comparison and show that the different experimental techniques capture results that are quantitatively different. This indicates the complexity of gas–solids flow in a riser due to the different length (from particle to large clusters) and timescales that must be captured and which usually interfere with the intrusive probes required to gather measurements in an opaque system. The overall trends of these experimental measurements are, however, qualitatively similar and can be used to partially validate our simulation predictions. At a height of 6.23 m above the gas distributor, the simulation results of W125 are in good agreement with the trends captured by the experimental data and show a large upward velocity at the center of the riser and a smaller, yet still upward, velocity near the wall. However, the results obtained with W1000 for the same restitution coefficient ($\text{W1000}-e=0.5$ in Fig. 23A) predict a flatter velocity distribution. These results are largely improved if a lower restitution coefficient ($e_{\text{cgp}}=0.14$, calculated according to Eq. (57) with $e_{\text{p}}=0.94$) is adopted in the simulation. Note that a simple calculation shows the number of computational parcels for this case of W1000 is only 1/8 of the W125 case. The flat profile resulting from a coarser simulation is explained by the lower frequency of collisions occurring as the number of parcels is reduced. At a height of 8.88 m, we notice, as was noted earlier, that the two different measuring technologies reveal quantitatively different results especially in the center region where the influence on the flow field of the FO is more pronounced than in the near-wall region. A simulation result similar to the lower height is observed, showing that simulations with smaller statistic

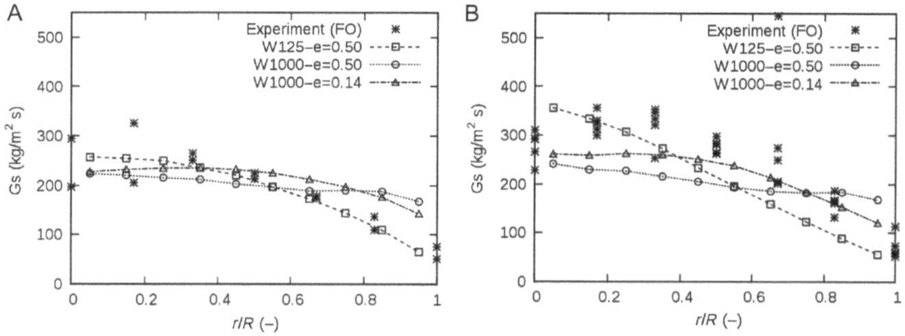

Fig. 24 Radial profiles of solid flux at different heights. (A) $H = 6.23\,\text{m}$, (B) $H = 8.88\,\text{m}$.

weight predict more accurate results and that reducing the restitution coefficient can help improve the accuracy of the coarse simulation.

Similar to the solids velocity profiles, the predicted radial profiles of time-averaged solids mass flux (Gs) at two different heights are compared with the available experimental measurements in Fig. 24. The time-averaged solids mass flux is the time-average of the instantaneous solids velocity by the instantaneous solids bulk density or simply the solids volume fraction since the particles' material density is not changing in this nonreacting flow (Benyahia et al., 2007). It is important to note that the solids flux may not be obtained by the product of time-averaged solids velocity and time-averaged solids bulk density because these two flow variables (velocity and volume fraction) are usually correlated in multiphase flows. This fact is intuitive and also experimentally measurable for riser flows because regions of high solids density are expected to have low or downward solids velocity, and high solids and gas velocity are expected in dilute regions. The existence of this obvious correlation between solids velocity and density makes it incorrect to measure the time-average of these two quantities, then assume the solids flux is equal to their product (Kirbas et al., 2007). Fig. 24 shows that smaller spatial gradients are measured for the solids mass flux than those seen previously for the solids velocity profiles. This indicates that although the solids velocity in the center of the riser is much higher than the near-wall region, the solids volume fraction in the near-wall region is much higher than in the center of the riser as commonly seen for this type of core-annulus flow regime. Again, the trends predicted by our simulation compare well with experimental data, although a great amount of uncertainty exists in the experimental data at a height of 6.23 m due simply to a repeat of experiments. At 8.88 m above the gas distributor, the profile of solids mass flux near the center is slightly underpredicted, especially for the largest parcel size, without corrected restitution coefficient (W1000 $-e=0.5$ in Fig. 24B). Similar to solids velocity predictions, the solids mass flux predictions show some improvement in the captured trends by lowering the solids restitution coefficient.

Generally speaking, the disagreements observed in the results for the solids velocity and mass flux can be explained by uncertainty associated with the experimentally measured solids flux as well as the fact that only the riser is simulated. Including the full CFB loop in our simulation could improve the numerical results, but this addition would introduce complexity in the geometry and would require additional computer resources. We should caution that experimental measurements of the transient solids circulation rate in this particular CFB indicated large-scale variations with large amplitudes, which will be difficult to capture numerically due to limited resources (Li et al., 2014). Other improvements to our simulation results may be obtained by refining the computational mesh for the fluid flow and using subgrid drag models that account for the unresolved effects of particle clusters smaller than the grid size. Although we showed it is possible to obtain improved predictions for large parcels, smaller parcels will always yield more accurate results. We view the use of parcels and their size as a compromise between computation speed and accuracy. The use of parcels inadvertently entrains numerical errors that can be quantified and may be acceptable to some modelers. The good news is that the CFD–CGHS is guaranteed to converge to the most accurate DEM solution as the parcel size is reduced to that of a particle. This means that uncertainties for this method can be quantified by comparing numerical results of the CGHS method with DEM. Obviously, this cannot be done in large-scale industrial reactors where DEM simulations are prohibitively expensive. In smaller systems, however, error estimation has showed that reasonably accurate results can be obtained for coarse-grained methods (Benyahia and Galvin, 2010). Thus, we suggest estimating these uncertainties in several smaller fluidized beds, then extrapolating to larger-scale systems. This is the only analysis missing in this study, and it will be conducted and published in the near future. It is important to note that such a study will involve a tremendous level of effort to estimate uncertainties associated with coarsening for different particulate flow regimes in different geometric scales. Nevertheless, such a study is important because it will provide the practitioner with a basic understanding of the errors associated with the model coarsening assumption for a particular application without having to conduct a Uncertainty quantification analysis from scratch—an endeavor which would be discouraging to most of us. In the next section, we will tackle the most important application of this coarse-grained method, industrial reactors. This and other methods are basically useless without real-world application to industrially relevant reacting systems. By conducting industrial applications, we hope that industrial practitioners and engineers will adopt and use this new technique, among their current tools, to understand, design, and troubleshoot their industrial reactors.

4. APPLICATION OF COARSE-GRAINED HARD SPHERE METHOD TO INDUSTRIAL REACTORS

The steps that we are taking to demonstrate the validity of the CGHS method are like those taken to develop a new reactor following an experimental approach. First, we have shown in this chapter that CGHS is fast and accurate for small-scale fluidized beds, and we then demonstrated the same in a pilot-scale cold-flow CFB. Now we will attempt to apply this method to larger reactors of interest to industry. These steps are necessary because of increased simulation complexity and resource requirements with increased system size. However, the most important reason to follow these steps in our simulations is because of the availability of detailed experimental data. Small fluidized beds tend to be the most instrumented with data made available by researchers in academia or national laboratories who are eager to publish their results. Data in large industrial fluidized beds, however, are much more difficult to obtain, and most data published by industry are made incomplete on purpose by not giving system geometry and/or operating conditions. We conduct industrial-scale simulations because some data are still available in the literature, as we will show in this section. Industrial applications are also important to attract potential users from industry who, we hope, will start using the techniques that we develop to make a more significant impact in real-world applications. Ultimately, we also hope that industry invests more resources on computational tools, such as CFD, to accelerate their development and also to guide such development to more practical applications of interest to engineering practitioners.

Classic chemical engineering processes such as those in oil refining and in petrochemical industrial processes are investigated here. A fluid catalytic cracking (FCC) regenerator unit used to regenerate spent catalyst is simulated in this section. A detailed experimental setup is publicly available for one regenerator process located in China with uniquely rich experimental data that cover flow dynamics, heat transfer for the exothermic reactions, and chemical concentrations of the products of coke combustion. The Methanol to Olefins (MTO) process that produces feedstocks for the petrochemical industry is also simulated with results compared with available experimental chemical conversion data. The last example that will be covered in this section relates to a rare-earth liquid–solids extraction process where experimental data are not yet available. In this last example, we show how the CFD model proposed in this chapter can be used for scaling up a counter-current fluidized bed process, which truly bridges the scales from laboratory to pilot to industrial size reactors. We hope that these examples will affirm the use of CFD in industry by those who already use it, and attract reluctant industrial practitioners to consider using CFD as an important tool for designing and troubleshooting, in addition to their preexisting arsenal of tools and expertise.

4.1 Simulation of a Fluidized Catalytic Cracking Regenerator

FCC is one of the most widely used and oldest conversion processes in refineries. A detailed description of the different FCC processes and their evolution in recent history is given in a previous study (Avidan and Shinnar, 1990) and only a very limited description of the regenerator process is provided here, for brevity. Through the FCC unit, heavy oils are converted into a variety of higher value light products like gasoline. The chemical reactions happen at the surface of the catalyst. A side effect of these catalytic cracking reactions is the deposited coke on the catalyst surface, which deactivates the catalyst and reduces the quality of the final products. Thus, a regenerator is designed to burn the deposited coke and restore catalytic activity. As shown in Fig. 25, the spent catalyst flows into the regenerator and deposited coke is burned out with air. Then, the regenerated catalyst flows back into the riser. The heat generated by burning coke in the regenerator will be used in the riser reactor to vaporize the heavy oil feed, which is usually fed as droplets dispersed through nozzles with the help of pressurized steam. The FCC system, which consists of the reactor and the regenerator, operates as a CFB where the catalyst is constantly circulating between the reactor and the regenerator. For the reactor part, the interested reader can find in our previous research (Lu et al., 2017e) coarse-grained simulation results of the flow field and catalyst residence time distributions in a pilot-scale FCC riser. A major assumption is made in this study to decouple the effects of the riser reactor and the catalyst regenerator so the latter can be simulated separately. This assumption is reasonable when the FCC unit is at steady-state

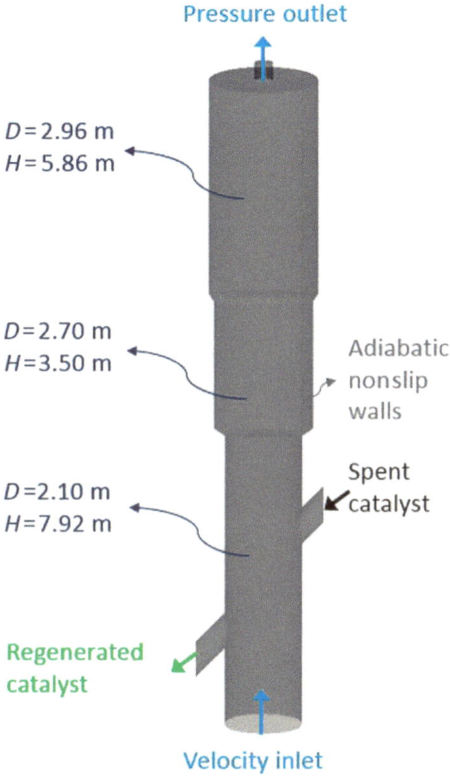

Fig. 25 Geometry and boundary conditions of simulated regenerator.

operation, where the catalyst circulation rate is relatively constant providing low fluctuations in the regenerator inlet conditions. The case studied here is the only one, to our knowledge, available in the open literature that provides detailed experimental data to be used to further validate the model proposed in this study. The most important aspect of modelling industrial-scale reactors is to provide the computation speed of our proposed CFD–TDHS simulations; engineering practitioners will appreciate the lightning-fast execution speed of the models developed in this study, which are conducted on very accessible computer resources. Thus, the regenerator in this study is simulated at an astonishing speed of 210 s/day using only 64 cores (4 computers with 16 cores each).

4.1.1 Simulation Setup

As shown in Fig. 25, a relatively small industrial FCC regenerator (70 thousand tons per year) is simulated in this study. The catalyst consists of solid

Table 6 Correlations for the Gas Mixture Properties

Property	**Corrections**
Density	$\rho_g = P/RT\sum \frac{Y_{gi}}{M_{gi}}$
Viscosity	$\mu_g = 1.672 \times 10^{-6}\sqrt{T} - 1.058 \times 10^{-5}$
Thermal conductivity	$\lambda_g = 5.526 \times 10^{-5}\sqrt{T} + 0.01155$
Heat capacity	$C_g = \sum Y_{gi}C_{gi}$
Molecular weight	$M_g = 1/\sum \frac{Y_{gi}}{M_{gi}}$

particles with a diameter of 60 μm and density of 1500 kg/m^3. The gas phase is a mixture of oxygen, nitrogen, carbon dioxide, carbon monoxide, and water vapor. The gas-phase properties are calculated using the ideal gas law and summarized in Table 6. The geometry of the regenerator is taken from a previous publication (Chang et al., 2013) and consists of a cylindrical fluidized bed with a diameter that varies between about 2 and 3 m and a height of more than 17 m. The details of the catalyst inlet and outlet pipelines were not provided in the publication. Also not provided were detailed information of the geometry transition between the main sections of the bed or the design of the top outlet. We estimated the spent catalyst inlet and the regenerated catalyst outlet pipe diameters as 0.7 m. The regenerated catalyst is only allowed to leave the system if the catalyst inventory in the system is greater than 4.95 tons. Effectively, a valve opens to let the catalyst leave the regenerator, but closes as the inventory in the system falls below 4.95 tons. In the simulation, the closing of this "valve" is achieved by setting the catalyst velocity to zero in the cell adjacent to the outlet represented at the bottom left side in Fig. 25.

The boundary conditions are set according to available industrial data. The spent catalyst flows into the regenerator at a rate of 22.7 kg/s with a temperature of 753 K. The carbon mass fraction in the spent catalyst is 0.9% and the hydrogen mass fraction is 0.072%. The inventory in the regenerator is initialized to 4.950 tons. The outlet pressure is set to a constant value of 140 kPa. For the main air inlet, we calculated that the air flows into the regenerator at a constant superficial velocity of 0.495 m/s, at a temperature taken from reference paper set to 573 K, and at a pressure taken from the referenced book (Lu and Wang, 2002) set to 160 kPa. According to industrial data, the coke content of the regenerated catalyst is about 0.39 wt%, while that of the spent catalyst is about 0.9 wt% with a hydrogen content of 0.072 wt%. In this simulation, we assume the mass ratio of carbon to hydrogen of regenerated and spent catalyst to be the same, thus the carbon and hydrogen on the regenerated catalyst is about 0.3611 and 0.0289 wt%, respectively. Based on the spent catalyst flow rate and the carbon conversion rate, the oxygen mass flow rate required to convert that carbon is about 0.4045 kg/s, which translates to an air flow rate of about 1.7435 kg/s. Thus, a mass balance on the coke contents of the catalyst was used to calculate the oxygen/air flow rate, which was then used, along with the set temperature/pressure, to calculate the inlet air gas velocity of 0.495 m/s, based on the ideal gas law. For heat transfer, the walls are considered adiabatic. All boundary conditions used in the simulation are summarized in Table 7, while those unavailable in the literature were estimated to conduct our simulation.

For initial conditions, the regenerator is separated into a dense phase up to a height of 7.9 m, while the top of the reactor consists of an initially empty freeboard. A total mass of 4.95 tons of regenerated catalyst particles are initialized in the dense phase of regenerator. The initial conditions are summarized in Table 8.

As discussed previously, the CGHS method is used to simulate this industrial-scale regenerator with a drag model proposed by Gao et al. (2009). The numerical parameters needed to simulate the fluid flow on a computational grid with Lagrangian coarse-grained particles are summarized in Table 9.

4.1.2 Chemical Reaction Kinetics Model

The coke deposited on the catalyst is burned in the regenerator to recover the catalyst activity. The main chemical reactions are carbon burning and hydrogen combustion. A general scheme of this reaction can be expressed as

$$\mathrm{CH}_{12n} + \left(\frac{r+0.5}{r+1} + 3n\right)\mathrm{O}_2 \rightarrow \frac{r}{r+1}\mathrm{CO}_2 + \frac{1}{r+1}\mathrm{CO} + 6n\mathrm{H}_2\mathrm{O}$$

where r denotes the molar ratio of carbon dioxide to carbon monoxide. In this research, we use the reaction rate provided by Chang et al. (2013). For carbon burning, the reaction rate is given as

$$R_\mathrm{C} = -\frac{dC}{dt} = k_\mathrm{c} P_{\mathrm{O}_2} C_\mathrm{c}$$

$$k_\mathrm{c} = 2.78 \times 10^8 e^{\frac{-161.2\times 10^3}{RT}}$$

$$r = 5.28 \times 10^{-3} e^{\frac{34860}{RT}}$$

Table 7 Boundary Conditions

Term	Simulation Value	Industrial Data
Bottom inlet air velocity (m/s)	0.495	NA (not available)
Bottom inlet air pressure (kPa)	160	160
Bottom inlet air temperature (K)	573	573
Bottom inlet air oxygen mass fraction	0.2320	0.2320
Bottom inlet air carbon dioxide mass fraction	0.0005	0.0005
Bottom inlet air carbon monoxide mass fraction	0.0000	0.0000
Bottom inlet air water vapor mass fraction	0.0000	0.0000
Bottom inlet air nitrogen mass fraction	0.7675	0.7675
Top outlet pressure (kPa)	140	140
Spent catalyst inlet air velocity (m/s)	0.5	NA (not available)
Spent catalyst inlet air pressure (kPa)	140	NA
Spent catalyst inlet air temperature (K)	735	NA
Spent catalyst inlet air oxygen mass fraction	0.0212	NA
Spent catalyst inlet air carbon dioxide mass fraction	0.2617	NA
Spent catalyst inlet air carbon monoxide mass fraction	0.0093	NA
Spent catalyst inlet air water vapor mass fraction	0.0416	NA
Spent catalyst inlet air nitrogen mass fraction	0.6662	NA
Spent catalyst mass flow rate (kg/s)	22.7	22.7
Spent catalyst carbon mass fraction	0.00900	0.00900
Spent catalyst hydrogen mass fraction	0.00072	0.00072
Spent catalyst temperature (K)	753	753
Spent catalyst inlet voidage	0.9	NA
Regenerated catalyst outlet pressure (kPa)	140	NA
Regenerated catalyst outlet solid mass flow rate (kg/s)	22.7	22.7
Walls momentum transfer	Nonslip	NA
Walls heat transfer	Adiabatic	NA
Walls species transfer	Zero flux	NA

Table 8 Initial Conditions

Term	Simulation Value	Industrial Data
Dense phase air velocity (m/s)	0.00	NA (not available)
Dense phase air pressure (kPa)	140	140
Dense phase air temperature (K)	912	912
Dense phase air oxygen mass fraction	0.0212	NA
Dense phase air carbon dioxide mass fraction	0.2617	NA
Dense phase air carbon monoxide mass fraction	0.0093	NA
Dense phase air water vapor mass fraction	0.0416	NA
Dense phase air nitrogen mass fraction	0.6662	NA
Dense phase catalyst carbon mass fraction	0.003611	NA
Dense phase catalyst hydrogen mass fraction	0.000289	NA
Dense phase catalyst temperature (K)	912	912
Dense phase catalyst volume fraction	0.8794	0.8794
Dilute phase catalyst volume fraction	0.0000	0.0050
Dilute phase air velocity (m/s)	0.0000	NA
Dilute phase air pressure (kPa)	140	140
Dilute phase air temperature (K)	930	930
Dilute phase air oxygen mass fraction	0.0212	NA
Dilute phase air carbon dioxide mass fraction	0.2617	NA
Dilute phase air carbon monoxide mass fraction	0.0093	NA
Dilute phase air water vapor mass fraction	0.0416	NA
Dilute phase air nitrogen mass fraction	0.6662	NA
Dilute phase catalyst carbon mass fraction	0.003611	NA
Dilute phase catalyst hydrogen mass fraction	0.000289	NA

For hydrogen combustion, the reaction rate is given as

$$R_{\mathrm{H}} = -\frac{dC_{\mathrm{H}}}{dt} = k_{\mathrm{H}} P_{\mathrm{O}_2} C_{\mathrm{H}}$$

$$k_{\mathrm{H}} = 4.17 \times 10^8 e^{\frac{-157.569 \times 10^3}{RT}}$$

Table 9 Numerical Parameters

Term	Simulation Value
Cartesian grid in x direction	15
Cartesian grid in y direction	90
Cartesian grid in z direction	15
Convection term discrete scheme	Superbee
Linear equation solver	BICGS
Hydrodynamic residual	1.0×10^{-3}
Temperature residual	1.0×10^{-3}
Species residual	1.0×10^{-3}
Gas phase time step (s)	Adaptive and no larger than 1.0×10^{-3}
Solid phase time step (s)	1.0×10^{-3}
Coarse grained particle diameter (m)	0.048
Particle–particle restitution coefficient	0.1
Particle–particle spring constant (kg/s$^{2)}$	1000
Particle–particle friction coefficient	0.1
Particle–wall restitution coefficient	0.1
Particle–wall spring constant (kg/s^2)	1000
Particle–wall friction coefficient	0.1

The heat of different reactions is provided below

$$C + O_2 \rightarrow CO_2, \quad Q = -393.51\,\text{kJ/mol}$$
$$C + \frac{1}{2}O_2 \rightarrow CO, \quad Q = -110.54\,\text{kJ/mol}$$
$$2H + \frac{1}{2}O_2 \rightarrow H_2O, \quad Q = -241.82\,\text{kJ/mol}$$

Since the reaction happens on the surface of the catalyst, the heat of the reactions is added as source terms to the particles heat balance. Physically, the heat of reaction is distributed between the catalyst and the gas with unknown proportions. This is not a major issue here because of the high rate of heat transfer between the particles and gas phase. Also, the simulation results of the temperature distribution show that the temperature difference between the gas phase and particles is less than 2 K. Since the timescale of the coke-burning reaction is very small compared with CFD simulation time steps, the reaction speed is scaled down by about six orders of magnitude to ensure the consumed mass of each species is less than 10% of the last step's value. This is done at the end of an old time step and before any CFD calculation is conducted at the new time step. This simplification avoids the additional computation cost of using a stiff chemistry solver, which employs a very small time step to resolve the chemical reaction process.

4.1.3 Results and Discussion

From the catalyst flow rate and inventory, we estimate the mean residence time to be about 218 s. Here, the simulation was conducted for 1000 s with the last 500 s used for time averaging the results to compare with industrial data. Such a long simulation time for CFD is only achieved because of the main coarsening assumption used in this current modelling approach. Some instantaneous snapshots of the numerical results are provided in this section to better understand the evolution of the gas particles reacting flow system.

4.1.3.1 Hydrodynamics of Fluidized Particles in the Regenerator

The simulation successfully captured the two-phase distribution of catalysts in the regenerator and compared quantitatively well with the available industrial data. This distribution is clearly shown in Fig. 26 where the bed density near the bottom of the regenerator is much larger than the top region. The large oscillation near 8.0 m is due to the diameter of the regenerator increasing from 2.1 to 2.7 m in this region, where the transition is not very smooth. This is intentional in the design of the regenerator where the sudden expansion of the geometry causes the gas and particles to slow down and, thus, lower the entrainment of catalyst into the freeboard. In fact, the simulation results in Fig. 26 indicate no catalyst entrainment in the top freeboard region of the bed. However, the industrial data in this top region do show the existence of a small but measurable quantity of catalyst that was not captured by the simulation due to the assumption of a single particle size used to represent the particle size distribution (PSD). This PSD was not provided by the industrial data and, so, was not considered in the simulation. We must emphasize the fact that small particles, or fines, in a PSD will

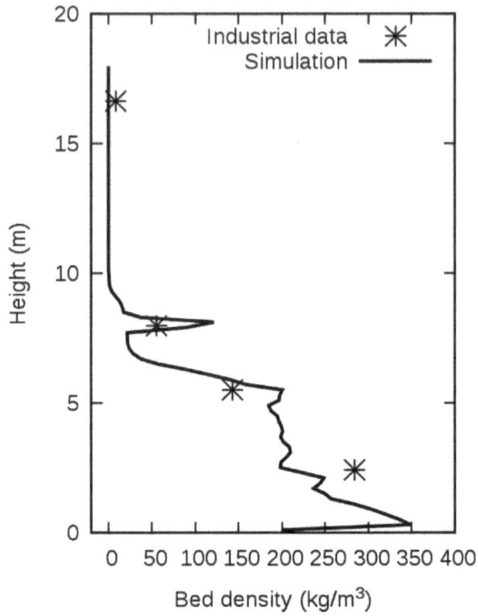

Fig. 26 Comparison of predicted (averaged from 500 to 1000 s) bed densities with industrial data.

generally be entrained to the freeboard region necessitating the use of cyclones (both primary and secondary). We assume that the lack of fines entrainment in our simulation contributed to some of the discrepancy between industrial data and numerical results.

4.1.3.2 Temperature Profiles and Heat Transfer

Through the burning of coke, the regenerator also acts as a heat exchanger to raise the temperature of the catalyst, thus increasing the cracking reaction speed in the riser. In this industrial operation, the inlet air temperature is about 573 K and increases significantly just after entering the regenerator because of the heat released from burning the coke. Fig. 27 shows the instant gas-phase temperature distribution in the middle plane of the regenerator (recall that the regenerator is a 3D cylindrical system). Initially, the dense phase is set to 912 K according to industrial data (see also Table 3). Just after this initialization of the temperature field, the temperature of the right side inlet decreased sharply due to the relatively lower temperature of the inlet spent catalyst. The first few snapshots early in the simulation show the initially high temperature set in the freeboard decreasing

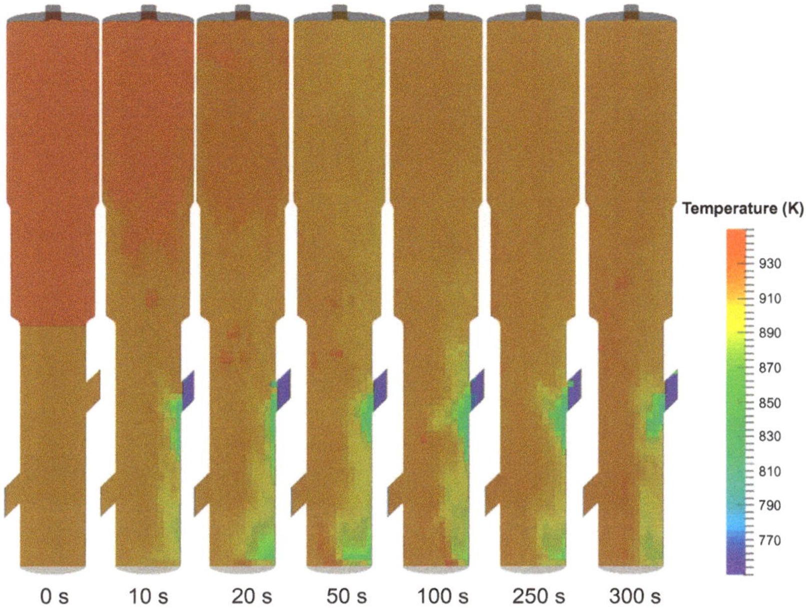

Fig. 27 Gas temperature distribution in the regenerator.

gradually. After about 100 s of simulation time, the temperature distribution in the regenerator is stabilized, with a lower temperature in the dense bed region at the bottom of the regenerator and a higher temperature in the dilute phase.

The same phenomenon can also be observed when examining the catalyst temperature, as shown in Fig. 28. The spent catalyst flows into the regenerator at a relatively lower temperature of 753 K. After entering the regenerator, most of the spent catalyst flows down to the bottom near the wall, bypassing most of the dense bed material. This is clearly designated in the figure by the blue color of the spent catalyst changing to green as it falls near the same side of the regenerator wall as the catalyst inlet. Near the bottom region, and due to the high oxygen concentration from the main inlet air, the burning reaction rate is much faster and the released heat quickly increases the temperature of these spent catalyst particles. The temperature of the regenerated catalyst exiting the system from the lower left pipe is similar to temperature of the catalyst near the regenerator freeboard, indicating a good overall mixing of catalyst in this fluidized bed regenerator. Fig. 29 shows the time-averaged gas-phase temperature distribution along

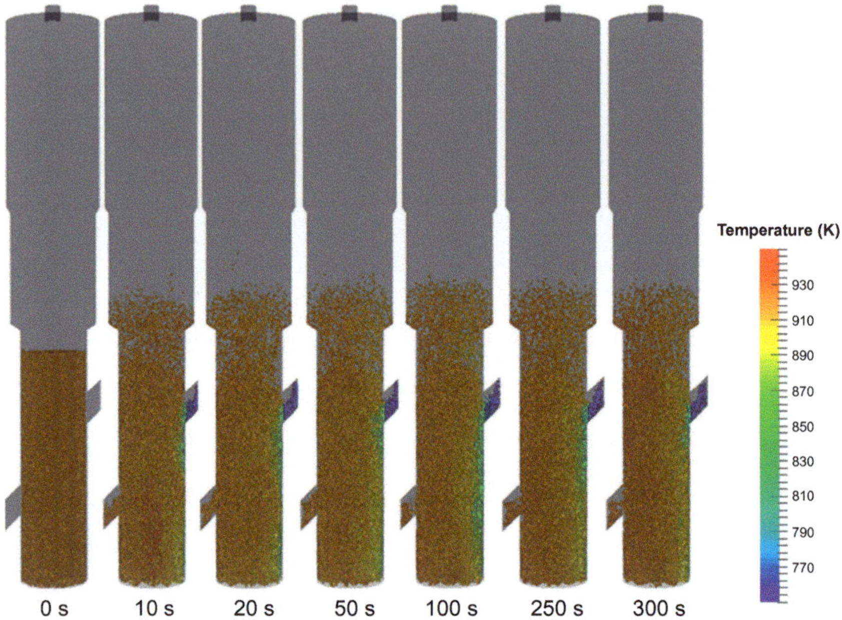

Fig. 28 Catalyst temperature distribution in the regenerator.

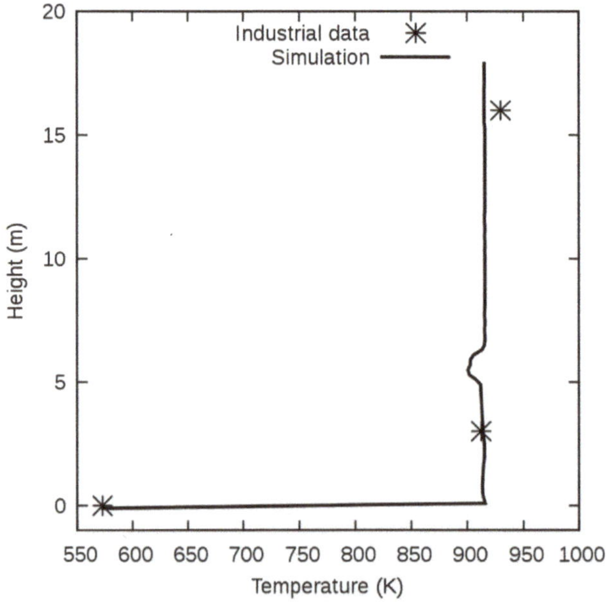

Fig. 29 Comparison of predicted temperature with industrial data.

the axial direction. Furthermore, these time-averaged results were also spatially averaged in the radial direction so that the final numerical data depended only on the height inside the reactor. It is clear that the temperature increases sharply at the bottom of the regenerator due to relatively cold air entering a very hot environment maintained by coke combustion reactions. In the bottom dense phase, the predicted temperature is about 912 K, which is same as industrial data. The small bump in the temperature profile at about the 5 m height is due to relatively cooler spent catalyst feed. At the top dilute phase, the predicted temperature is about 920 K, while the industrial data show about 930 K. Contributing to this difference are many factors, mostly related to the idealization of the model to represent the physical reality, and the most important is the lack of particle entrainment caused by a lack of PSD consideration. The errors in the simulation using a single particle size not only cause discrepancies with the bed density measurements but also, as seen here, with temperature predictions. Nevertheless, the error is only slightly more than 1.0%, which is acceptable for most engineering applications. Thus, it is reasonable to conclude that the simulation captures reasonably well the entire temperature distribution in the regenerator.

4.1.3.3 Species Concentration in the Regenerator

According to industrial data, about 4.95 tons of regenerated catalyst particles are initialized in the dense phase of the regenerator with an initial coke concentration of the catalyst set to 0.39%. This initial concentration of coke can burn quickly and completely if the combustion reactions are not controlled because fresh air with a high concentration of oxygen flows into the regenerator from the main air inlet at the bottom of the reactor. Thus, the reaction kinetics are modified by simply setting the reaction rate to zero if the carbon concentration is less than 0.39%. From an industrial point of view, burning this remaining coke concentration is not possible, because it is buried inside the catalyst and not easily accessible to the highly reacting oxygen. Therefore, simply stopping the reaction at this level of coke concentration is reasonable and practical for the purpose of easily conducting our CFD simulation.

Fig. 30 shows the detailed carbon mass fraction and particle flow field in the regenerator. In the dilute region (as shown in Fig. 30A), the particle velocity, as indicated by the velocity vectors attached to the parcels, is randomly distributed with a low concentration of carbon; however, some particles with high carbon concentration are still visible in this figure. At the spent catalyst inlet region (as shown in Fig. 30B), most of the spent

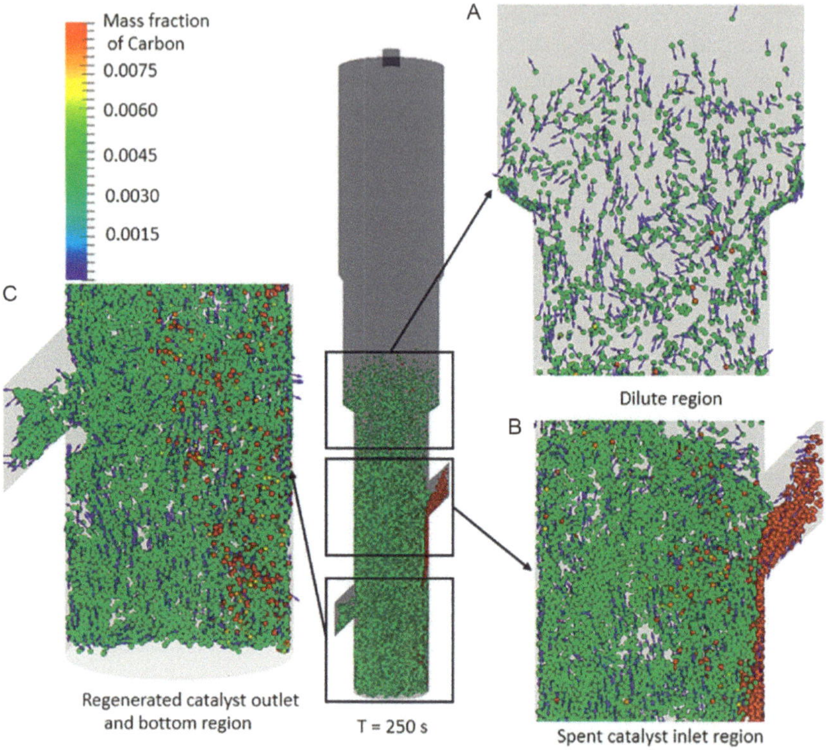

Fig. 30 Carbon mass fraction and particle flow field in the regenerator.

catalyst particles flow down toward the bottom near the wall, bypassing most of the dense bed material. Still, fewer particles manage to diffuse into the center of the regenerator and flow upward toward the dilute region. This explains why some particles with high carbon concentration are found in the dilute region. At the regenerated catalyst outlet and bottom region (as shown in Fig. 30C), the cross-sectional plot shows bubbles (voids) surrounded by a densely concentrated bed of particles. Particles generally flow down near the walls until reaching the bottom uniform air distributor where a flow reversal occurs. It is clear that most spent catalyst flows down counter-currently on the right side of the wall against the upward flowing gas, then moves cocurrently with the gas at the core of the bed, and finally reverses flow near the top of the bed and flows downward on the left side of the wall to leave the regenerator through the regenerated catalyst outlet. This complex flow behavior is easily seen in the plot by observing changes to velocity vectors attached to the parcels, while the

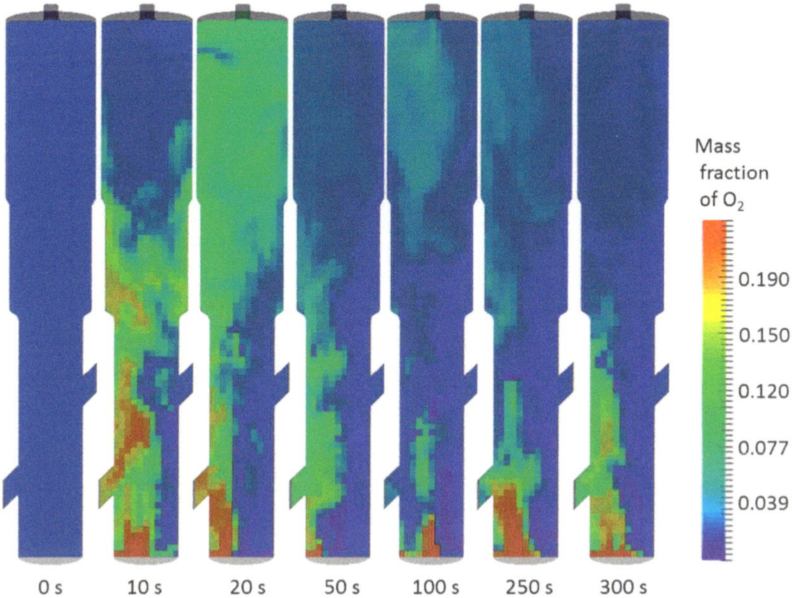

Fig. 31 Oxygen mass fraction distribution in the regenerator.

carbon concentration acts as a tracer that helps follow the flow of fresh particles in the regenerator.

Fig. 31 shows the oxygen mass fraction distribution in the regenerator. Initially, the oxygen mass fraction in the dilute phase of the regenerator is set to 0.0212, according to industrial data, near the top outlet. For the dense phase, the industrial data are not available and were initialized to the same value as in the dilute phase. The effect of using this value is relatively unimportant since the residence time of gas phase is only about 20 s, which is much smaller than the total simulation time. At 10 s, the oxygen flows from the bottom air distributor to the top region with the right part of the bottom region depleted of oxygen by carbon burning reaction because of the preferential flow of spent catalyst in that specific region of the reactor. At 20 s, the mass fraction of oxygen in the top dilute region is much higher than initial value because of the inexistence of reacting particles in the freeboard region. At 50 s, the mass fraction of oxygen in the top dilute region reduces to a very small value. This is because oxygen, which previously accumulated from the initial conditions, flows out from the top outlet and the newly added oxygen is now consumed at the bottom dense region due to the presence of reactive particles. After 250 s, the system stabilizes and reaches a near steady-state operation.

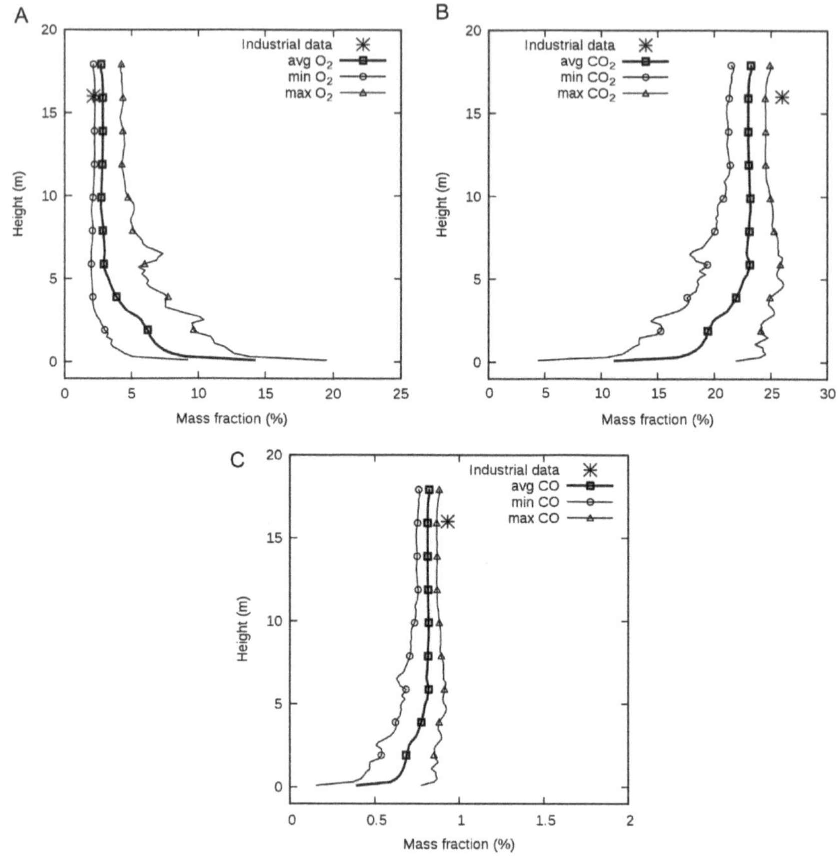

Fig. 32 Time-averaged oxygen (A), carbon dioxide (B), and carbon monoxide (C) mass fraction distributions along the height of the regenerator.

Fig. 32 shows the time-averaged oxygen, carbon dioxide, and carbon monoxide mass fraction distributions in the regenerator where the numerical data were also space-averaged along all radial positions so that there is a dependence on height only. The data were time-averaged using only the last 500 s of simulation from 500 to 1000 s as previously mentioned. Since it is not known how or where exactly the experimental data were taken, we decided to plot the minimum, maximum, and averaged values of these chemical species mass fractions along the height of the reactor. The predicted values compare well with available industrial data. In the dense region, the oscillation of mass concentration is larger than that in the dilute region. This is because, in the dense region, the fluidized bed is in turbulent fluidization regime, which leads to a highly dynamic flow field. A longer time-averaging period will reduce such oscillations and usually results in smoother profiles. For carbon dioxide, the predicted value of the mass fraction at the dilute phase is about 23%, while the industrial data are about 26% leading to a relative error of about 11%. For carbon monoxide, the predicted mass fraction value at the dilute phase is about 0.82%, while the industrial data are about 0.93% leading to a relative error of about 12%. Notice that the concentration of carbon dioxide and carbon monoxide mass fractions was not plotted in the same way as oxygen in Fig. 31. This is because of the similarity of these plots where the concentrations of carbon dioxide and carbon monoxide can easily be obtained from that of oxygen by simply reversing the color plots (blue becomes red and vice versa). Thus, we chose not to plot these additional contour profiles for the sake of brevity.

The simulation of the FCC regenerator process shows that our newly developed CFD–TDHS approach can easily simulate this relatively complex reactor. The ease of simulation stems from two important facts: (1) the simulation's extremely fast execution speed, as mentioned earlier, is crucial to obtaining meaningful results in just few days; (2) this method is reasonably accurate even though the simulation is relatively coarse, as indicated by the size of the parcels, compared to our previous small-scale simulations. In fact, the use of numerical parcels that contain a large number of physical particles is the only reason why conducting such simulations is even possible. To obtain more accurate simulation results, the interested researcher can simply lower the coarseness of the simulation by including fewer particles in a parcel. As mentioned previously, the model errors associated with the main assumption of coarsening particles will be systematically studied in the near future.

4.2 Simulation of a Methanol to Olefins Reactor

MTO is an autocatalytic reaction that converts coal or natural gas-based methanol to valuable olefins, which are widely used in petrochemicals and relevant downstream processes (Tian et al., 2015). The design of MTO units is similar to the FCC process, where an online catalyst regenerator is operated with a main reactor. The difference is that the MTO reactor is operated under a bubbling or turbulent fluidization regime with a catalyst residence time much larger than the modern FCC system. This leads to a substantial computation cost to simulate the MTO reactor. In this section, a pilot-scale MTO reactor, where experimental data are available, is simulated with CFD–CGHS. To estimate the coke fraction on the catalyst, the system is first simulated with 0D simulation and then with 3D simulation. While a full 3D simulation tries to mimic the details of the actual system geometry, a 0D simulation is what chemical engineers usually call a perfectly mixed continuous stirred tank reactor where spatial gradients of flow variables (such as velocity, pressure, etc.) are nonexistent. We traditionally use 0D simulations to quickly verify/debug code written for complex chemical reaction kinetics without the addition of complex flow dynamics. Here, however, 0D simulation results can be used to obtain a good initial guess for 3D simulations as well as for comparison with 3D results to assess the effects of the flow dynamics on the reactor's overall performance. As far as the computation speed is concerned, the full 3D simulation runs at about 122 s/day using 128 cores (8 computers with 16 cores each), while the simplified 3D simulation (explained in the next section) runs at about 2200 s/day using 16 cores on a single computer. As mentioned previously in the FCC regenerator section, the speed of execution of these simulations is this method's main attribute, which we think is key to solving industrial problems.

4.2.1 Simulation Setup

The simulation setup of an MTO reactor is shown in Fig. 33A. The solid arrow lines indicate the flow directions of the gas phase, and the dash arrow lines show the flow directions of the catalyst. In this reactor, the main reactants, consisting of methanol and steam, flow into the reactor from the bottom distributor with a total mass flow rate of 27 kg/h with the mass ratio of methanol to steam set to 2:1. Another stream of steam enters the reactor with a flow rate of 0.12 kg/h from a lift tube connected to the bottom of the reactor. The catalyst is added into the reactor through a pipeline with a flow rate of 8.8 kg/h. These catalytic reactions produce coke deposits on the surface of the catalyst reducing its activity. The spent catalyst is then discharged through a central orifice on the distributor to the lifting tube and then transported to the regenerator. To maintain the same bed height as in the experiments, the inventory of the catalyst in the reactor is set to 9.0 kg and maintained by controlling the spent catalyst discharge rate using the same method discussed previously in the simulation of the FCC regenerator. The actual length of the lifting tube is 3.35 m, while only the top 0.05 m is simulated to save computer time. The EMMS drag model, described in the methods section, has been validated in previous TFM simulation study (Lu et al., 2016b) of the same reactor and is used in this simulation as well. Other simulation parameters are listed in Table 10.

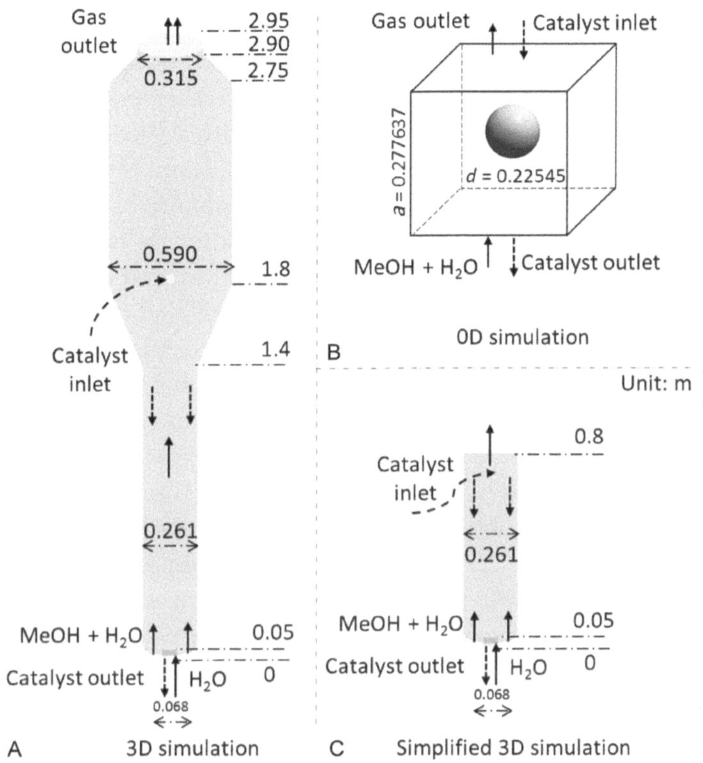

Fig. 33 Simulation setups of the MTO reactor.

Table 10 Simulation Parameters of the MTO Reactor

Properties	**Values**
Particle diameter, d_p (mm)	0.097
Parcel diameter, d_{CGP} (mm)	4.85, 9.7
Particle density, ρ_p (kg/m^3)	1500
Coke mass fraction in inlet catalyst	0.02
Restitution coefficient, e	0.1
Friction coefficient, μ	0.1
Spring constant, k_n (N/m)	100
Gas density, ρ_g (kg/m^3)	0.4288
Gas Viscosity, μ_g (Pas)	2.43×10^{-5}
CFD grid size, d_x, d_y, d_z (mm)	20, 20, 20
Gas phase time step (s)	5×10^{-3}
Solid phase time step (s)	1×10^{-4}

In this reactor, the residence time for the catalyst is more than 1 h, while the gas phase can flow through the reactor in several seconds. To save computational resources, the reactor is initialized with 9.0 kg of catalyst in the bottom region of the bed. With this initial condition, the hydrodynamics of this reactor can reach a stable state within a minute of simulation time; however, the coke deposit on catalyst requires a much longer time due to the slow reaction kinetics of coke production. Thus, and to save computational resources, we first carried out a cold full 3D simulation, shown in Fig. 33A, with a focus only on the hydrodynamics. After that, the volume of the main reaction region and averaged solid volume fraction is estimated based on the bed height obtained from this simulation result. Then, a virtual 0D reactor with only one CFD cell is constructed, as shown in Fig. 33B, and is assumed to be a cube with a length of 0.277637 m (this value will be explained later) where all the solid particles are lumped into a single parcel with a diameter of 0.22545 m. The inlet gas flow rate, catalyst flow rate, and species fractions are same as those used in full 3D simulation. The outlet flow rate is also the same as the inlet flow rate to maintain a constant mass in the reactor. However, the species fractions at the outflow are set to the same value of the only computational cell in this 0D reactor. The gas and particle velocities are set to zero and only the species transfer equations are solved. This 0D simulation can estimate the averaged coke deposit on catalyst quickly and was used previously for this purpose (Lu et al., 2016b). However, this simple 0D simulation cannot provide the coke deposit on different catalyst particles, which is a good metric to measure the mixing state of the catalyst in the reactor. To eliminate some of the major assumptions in the 0D simulation while still maintaining an acceptable computation speed, the reactor is also simulated with a simplified geometry as shown in Fig. 33C. In this simplified 3D simulation, only the bottom part of the main reactor is simulated. Since most of the particles are in this region, the uncertainties introduced through this simplification are very small compared to errors introduced in the 0D simulation. The major advantage lies in the computation speed of this simplified geometry, which will be much faster than the full 3D geometry due to the drastic reduction in the number of CFD cells.

4.2.2 Chemical Reaction Kinetics Model

The lumped reaction kinetics for MTO is used (Lu et al., 2016b). In this reaction model, seven parallel reactions are considered and the formation

Table 11 MTO Reaction Constant

Components	Rate Constant, k_i	Parameter, α_i
CH_4	0.00300	0.0649
C_2H_4	0.12463	0.1008
C_3H_6	0.16212	0.1996
C_3H_8	0.03467	0.4060
C_4H_8	0.07910	0.2924
C_5H_{10}	0.04895	0.3495
Coke	0.05859	0.38586

rate of each lump, R_i ($i = CH_4$, C_2H_4, C_3H_6, C_3H_8, C_4H_8, C_5H_{10}, and coke) is given by

$$R_i = n_i k_i \varphi_i C_{\mathrm{MeOH}} M_i$$

where n_i is the stoichiometric number; C_{MeOH} is the methanol concentration with a unit of mol/L; M is the molar weight (g/mol), which for coke is set to 84 g/mol; k_i is the reaction rate constant (L/g_{cat}/s); and φ_i is a deactivation function defined by

$$\varphi_i = \frac{1}{1 + 9e^{2(w_{\mathrm{coke}} - 7.8)}} e^{-\alpha_i w_{\mathrm{coke}}}$$

w_{coke} is the percentage of coke in the catalyst with a unit of g/100 g_{cat}. The rate constant k_i and α_i are listed in Table 11.

4.2.3 Results and Discussion

Fig. 34A shows the time-averaged gas volume fraction in the whole reactor. Most of the particles are in the bottom section of the reactor occupying a height of about 0.4 m from the gas inlet distributor. The top regions of the reactor, including the top sedimentation section, are almost empty. Based on this observation, the volume of the 0D reactor is set to 0.021401 m^3; in the simplified 3D simulation only a reactor of double this height (0.8 m) is simulated. Fig. 34B shows the instant particle distribution colored by vertical velocity with a 90% transparency to allow the observation of the flow structures that are otherwise invisible inside this opaque system. This figure indicates a dense region present near the bottom distributor and a relatively dilute turbulent region with clusters and bubbles above that. Fig. 34C shows the particle flow field, or flow patterns, at the middle plane

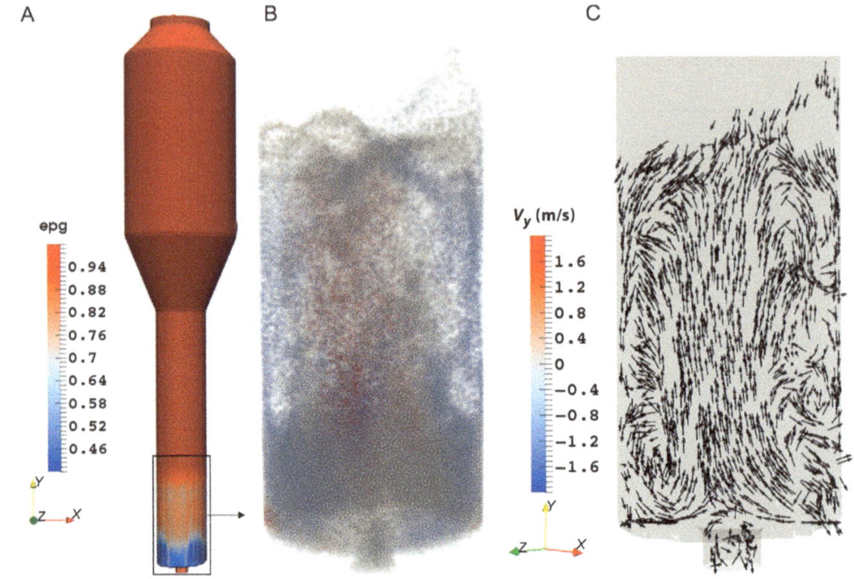

Fig. 34 Time-averaged void fraction (A), instant particle distribution (B), and flow field (C) in MTO reactor.

along the Z direction. The general flow patterns of the solids catalyst are as follows: the fresh catalyst is dropped into the top region of the fluidized bed and collides with the upward moving particles and gas. Then, the catalyst is transported to the near-wall region by the agitating effect of bubbles and moves downward to the bottom dense phase. In this dense region, the catalyst contacts the inlet gas and is dragged to the center of the bed. Through the significant agitation effects of bubbles inducing large-scale turbulence, the particles are well mixed, and this explains why we can attempt to use 0D simulation results to cheaply and accurately estimate the coke deposit.

After the initial hydrodynamic investigations presented earlier, the reactor was simulated for 8 h of flow time with chemical reactions using the simplified 0D geometry. Since the coke mass fraction on the regenerated catalyst is 2.0% (from industrial data), the 0D simulation is carried out with initial coke mass fraction of 2.0%. As shown in Fig. 35, the predicted coke accumulation result obtained with the 0D simulation is about 5.6%, which is slightly less than experimental data of about 6.0 wt%. However, the simulation speed in this simplified case is extremely fast, and the computation results are obtained within just few minutes on a single processor. This indicates that the 0D simulation method can be used as a tool to quickly and cheaply estimate equilibrium or steady-state chemical species concentrations

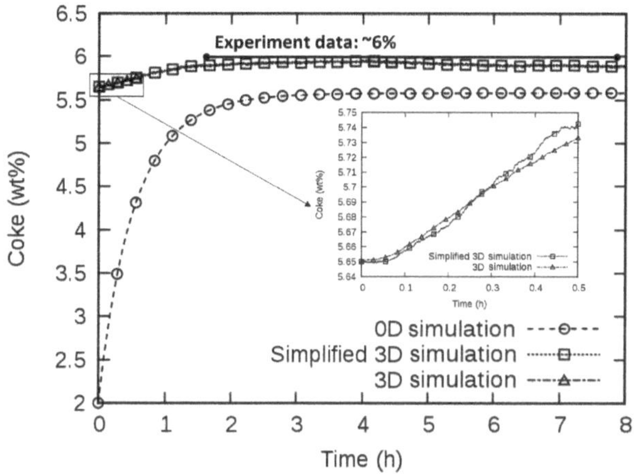

Fig. 35 Spatial-averaged coke mass fraction in MTO reactor.

in this bubbling fluidized bed. Based on the 0D simulation results, the initial coke mass fraction for 3D simulations is set to 5.65%, which is in the range of the 0D simulation and experimental data. Because of the significant computational requirements, the full 3D simulation is only carried out for 0.5 h, while the simplified 3D simulation is carried out for the full 8 h of total flow time. The results indicate that the simplified 3D simulation predicts a very similar profile to that obtained with the full 3D simulation (at least in the first 0.5 h), indicating that this simplification to the full 3D geometry is reasonable. The predicted coke mass fraction of the simplified 3D simulation is about 5.90%, which is in better agreement with experiment data compared to the 0D simulation.

Apart from the averaged coke mass fraction, the 3D simulation can also reveal detailed coke fraction statistics on the catalyst particles. Fig. 36 compares the coke fraction distributions and cumulative distributions at different simulation flow times. The initial trend shows a coke mass fraction value where a sharp change, or discontinuity, in the cumulative distribution occurs, indicating a large fraction of catalyst with a coke mass fraction equal to this value. After 0.5 h of simulation time, this discontinuity occurs at about 0.4, which represents 40% of newly added regenerated catalyst, since any catalyst with a coke concentration less than 5.65% was freshly added from the inlet rather than initially present in the reactor. As the simulation progresses, the curve becomes smoother, which means that most of the initial catalyst was discharged from the reactor. After 8.0 h of simulation time, even the distribution curve becomes very smooth with a maximum frequency occurring at about 7.5% of the coke concentration.

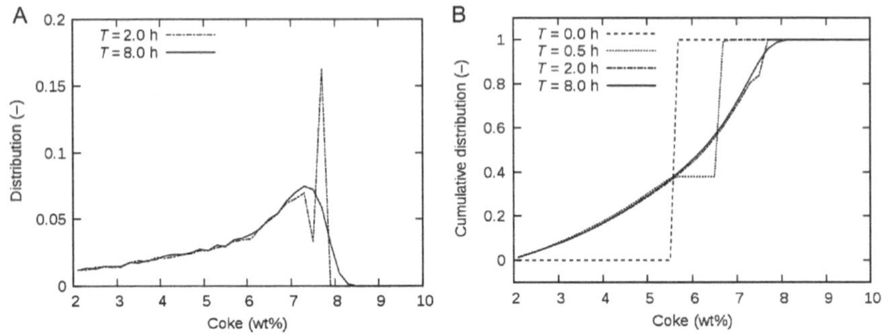

Fig. 36 Coke mass fraction distributions in the MTO reactor at different times. (A) Coke distribution and (B) coke cumulative distribution.

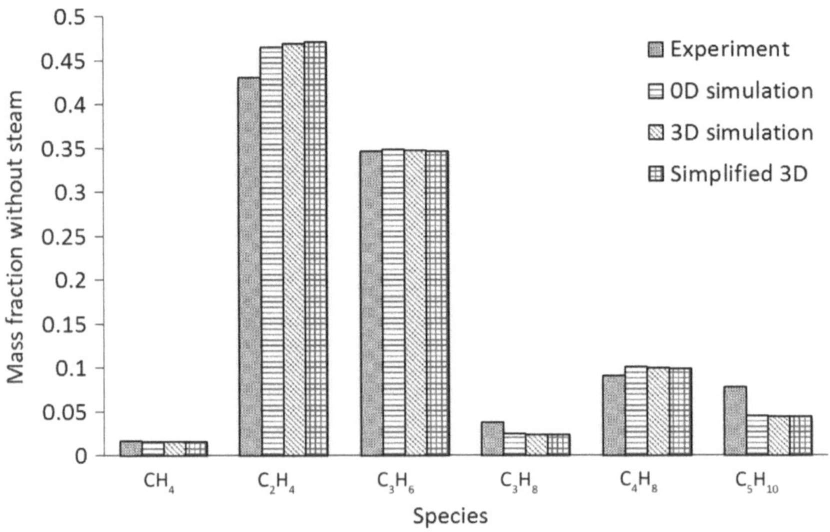

Fig. 37 Species mass fraction of the MTO process.

The mass fractions of different species in the products measured at the exit of the reactor are compared in Fig. 37. This figure indicates that the simulations using different assumptions can reasonably capture the experimental findings except for a slight over prediction of ethylene and underprediction of C_5H_{10}. The simulated light olefins (C_2H_4 and C_3H_6) account for about 81.2% of the total products, while the experimental results show a value of about 77.6%. This indicates the high selectivity of the catalyst used in MTO reactions.

The simulations of the two industrial reactors for FCC and MTO reactions conducted so far indicate that the CFD–CGHS method is efficient and reliable for these types of gas–solids reactors. Next, we will tackle a different type of multiphase flow reactor that deals with liquid–solids reactions. Unlike previous reactions, this liquid–solids flow involves an aqueous ion exchange reaction that will demonstrate the value of our newly developed technique to the widest possible industrial applications.

4.3 Simulation of a Rare Earth Elements Leaching Reactor

Rare earth elements (REEs) are widely used in high-technology products such as catalysts, lasers, cell phones, medical devices, and other applications. The first step of REE production is extracting the valuable REEs from ores, followed by a solution concentration and crystallization step and other purification steps. The traditional in-situ leaching of ores leads to severe environmental damages. A project we are currently working on involves the extraction of REE from coal by-products, such as clays from coal excavation and ash from coal burning/gasification, to minimize the environmental effects of coal mining and maximize its profits. A new counter-current leaching reactor is virtually designed using CFD to provide an environmentally friendly way to produce REEs. The leaching reaction is basically an ion exchange reaction, and the chemical reaction kinetics used in this research are taken from a published paper (Xiao et al., 2015). In this section, the implemented chemical reaction model is first validated against published experiment data using a small bubbling fluidized bed. After that, a small lab-scale counter-current reactor is simulated under different operating conditions to optimize REE production. Finally, this laboratory-scale reactor is virtually scaled up to both pilot- and industrial-scale reactors. The computation speed of the simulation conducted for the largest industrial-scale reactor is about 2700 s/day on a single computer node with 16 cores. This proves the ability of this newly developed technique to tackle not only industrial-scale problems but also their optimization using a rigorous CFD approach.

4.3.1 Validation of Chemical Reactions

The leaching process to extract ion exchangeable REE can be carried out with an acidic reagent. For example, solid phase REE is extracted from

REO (rare Earth ore) with magnesium sulfate solutions used as lixiviation agent. The chemical reaction mechanism is as follows (Xiao et al., 2015):

$$\begin{aligned}&\{Al_4[Si_4O_{10}](OH)_8\}_m^- \cdot 2nRE_{(s)}^{3+} + 3nMg_{(f)}^{2+}\\&\rightarrow \{Al_4[Si_4O_{10}](OH)_8\}_m^- \cdot 3nMg_{(s)}^{2+} + 2nRE_{(f)}^{3+}\end{aligned} \tag{102}$$

where s and f are solids and fluid phases, respectively. In this reaction, the aqueous phase cations replace REE sites with the REE extracted from REO and dissolved into an aqueous solution. The REE leaching with the ion exchange reaction can be explained as multiple steps using the shrinking core model. First, the reactive cations in the aqueous solution approach the REO surface by diffusion through a liquid boundary layer surrounding the REO particles. After the ions reach the surface of the REO, they diffuse into the porous solid particle until reaching the REE sites. Here, the ion migration inside the solid phase is called inner diffusion, while outer diffusion occurs just outside the particle surface. Once the cations encounter the REE, the ion exchange reaction occurs and the ammonium ions combine with the hosting minerals. During the ion exchange reaction, the REE core shrinks, while the REO particle maintains its initial spherical shape. We assume in this study that the REO particle shape and density remain constant during the ion exchange due to the low concentration of REE, which is assumed to be in the order of 1000 ppm. Finally, the extracted REE ions diffuse out to the acidic reagent. Here, the multiple step reactions are specified as (1) outer diffusion, (2) inner diffusion of ammonium, and (3) ion exchange reaction. Generally, and also proved by experimental data (Jun et al., 2010; Xiao et al., 2015), it is known that the reaction speed of ion exchange is faster than the inner diffusion phenomenon during REE extraction. Therefore, it can be safely assumed that the reaction rate is mainly controlled by the diffusion speed of reactive ions. Thus, the inner diffusion controlled shrinking core model was demonstrated to best describe the reaction mechanism by the following equations (Xiao et al., 2015):

$$1-\frac{2}{3}\alpha-(1-\alpha)^{2/3}=kt \tag{103}$$

$$\alpha=1-\left(\frac{r_c}{R_p}\right)^3 \tag{104}$$

$$k_{exp}(1/\min)=\frac{0.011}{R_p(mm)^{1.217}}\exp\left(-\frac{9480}{RT}\right) \tag{105}$$

where α is the leached fraction, k_{exp} is the chemical reaction rate from experimental measurements, r_c is the radius of the shrinking core, R_p is the constant radius of REO particle and R is the universal gas constant, and T is the reaction absolute temperature. In this model, the influence of Mg^{2+} concentration must be considered and the reaction rate is calculated as

$$k(1/\min) = \frac{C_{Mg}}{C_{Mg,\,exp}} k_{exp} \tag{106}$$

where $C_{Mg,exp}$ and C_{Mg} are the Mg^{2+} concentration in experiment and simulation, respectively. Since the experiment is carried in a fast stirred 500 mL container with ample $MgSO_4$ solution, the system is considered perfectly mixed and the Mg^{2+} concentration is assumed constant.

To validate our implementation of the reaction kinetic model in MFIX, the REE extraction process is simulated in a small liquid–solids fluidized bed. The simulation parameters are listed in Table 12. In the experiment (Xiao et al., 2015), the particle size is in the range of 0.45–0.60 mm. Since the detailed PSD is not given, a single particle size is used in each simulation.

Table 12 Simulation Parameters of Small Liquid–Solids Fluidized Bed

Properties	**Values**
Particle diameter, d_p (mm)	0.45, 0.6
Statistic weight, W	250, 500, 1000
Particle density, ρ_p (g/cm^3)	1.34
REE mass fraction in particle	0.00102
NH_4 mass fraction in particle	0.0
Ash mass fraction in particle	0.99898
Restitution coefficient, e	0.7
Friction coefficient, μ	0.1
Liquid density, ρ_l (g/cm^3)	1.0
Viscosity, μ_l (g/cm/s)	0.01
REE mass fraction in liquid	0.0
NH_4 mass fraction in liquid	0.002730
SO_4 mass fraction in liquid	0.007346
H_2O mass fraction in liquid	0.989924
Fluidized bed dimensions, L_x, L_y, L_z (mm)	80, 320, 80
CFD grid size, d_x, d_y, d_z (mm)	10, 10, 10

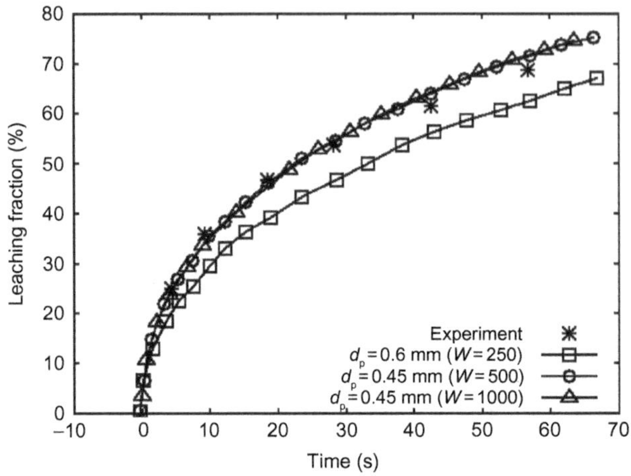

Fig. 38 Comparison of REE leaching fraction obtained experimentally and with simulation.

In these simulations, the Wen—Yu drag model, described in the methods section, is used. Also, to investigate the influence of parcel size on simulation results, different statistic weights are adopted in our calculations.

Fig. 38 compares the REE leaching fraction between experimental findings and simulation results using different statistic weights and particle diameters. Although the experimental data are from a continuously stirred three-necked flask, the simulation results compare well with experimental data because a fluidized bed is a fully mixed reactor, especially for a homogeneous liquid–solids flow that is unlike the heterogeneous gas–solids flow systems studied previously in this chapter. The results are also insensitive to different statistic weights considered in this study. This proves that the implemented chemical reaction mechanism can accurately capture the REE leaching reaction. Also, the simulation with a diameter of 0.45 mm compares well with experimental data. This indicates that, in the experiment, most of the particles have a diameter of 0.45 mm.

4.3.2 Simulation of a Small Counter-Current Reactor Under Different Operating Conditions

Recent research (Moldoveanu and Papangelakis, 2016) shows that column leaching provides a more efficient alternative to the batch process and constitutes an important step toward simulating the in-situ leaching of REEs. However, this system is limited by its liquid flow rate through a densely packed bed. Kwauk (1979, 1992) investigated a leaching process in a counter-current fluidized bed. The particles are added from the top of the reactor and fall counter-currently against a rising stream of liquid into the fluidized leaching section. Fresh leaching liquid enters the system through special sprayers located near the bottom of the reactor. Below the liquid inlet region, the solids are discharged from the bottom with an automatically controlled valve. In our simulation, the inlet flow is simulated with point sources to avoid considering complex inlet designs, and the closing and opening of the valve are controlled by the total number of particles in the reactor, as done in the two industrial cases from the previous sections. A counter-current fluidized bed capable of leaching REEs shown in Fig. 39 was simulated using coarse-grained CFD–DEM with different operating conditions, listed in Table 13. As in the

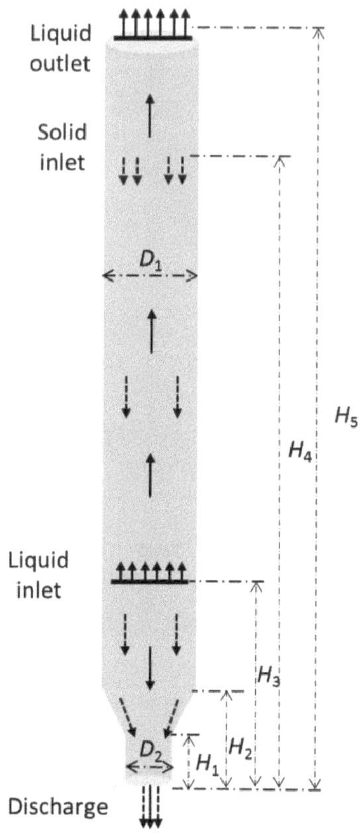

Fig. 39 Schematic of counter-current leaching reactor. *Modified from Lu L, Yoo K, Benyahia S: Coarse-grained-particle method for simulation of liquid–solids reacting flows.* Ind Eng Chem Res *55:10477–10491, 2016a with permission for the reprinted content. Copyright (2016) American Chemical Society.*

Table 13 Simulation Parameters of Counter-Current Leaching Reactor

Case	Base	Recycle50	L36	L36S36
Liquid flow rate, L (kg/h)	72	72	36	36
Solid flow rate, S (kg/h)	18.41	18.41	18.41	36.82
Liquid recycle fraction (%)	20	50	20	20

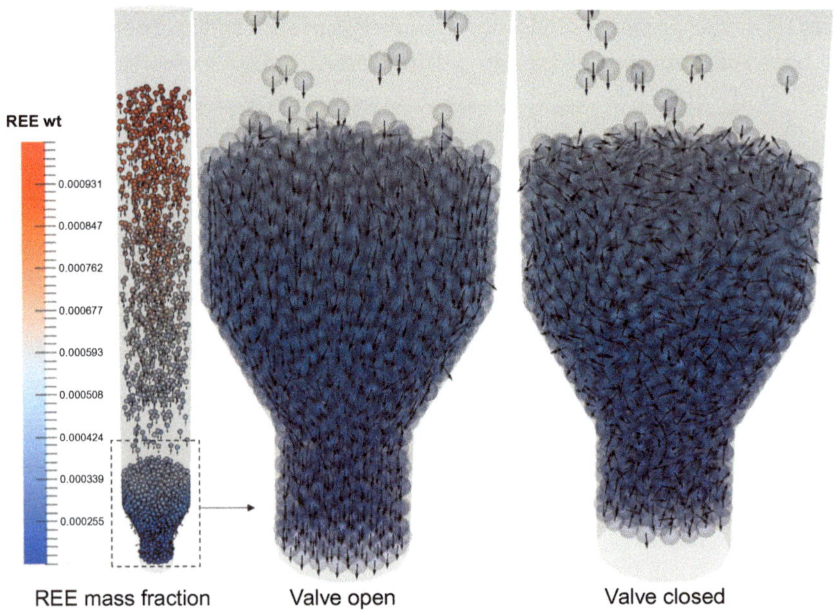

Fig. 40 REE mass fraction and particle flow field.

previous simulation, the Wen—Yu drag model is also used here. Other parameters are the same as the previous simulation of fluidized bed in Table 12.

Fig. 40 shows several snapshots of the simulated results and reveals the flow behavior in different sections of this counter-current reactor. It shows that most of the REEs are extracted out as the particles are moving toward the bottom of the reactor. Similar to Kwauk's experimental observation (Kwauk, 1979, 1992), the reactor has a top dilute region and bottom dense region. When the inventory of particles in the reactor is larger than a user-defined value, the bottom valve will open and the leached particles can be discharged. After this discharge, the valve is closed again by setting the particles' velocities to zero in the adjacent cells to the outlet patch. A complex flow field showing particles agitation is observed when the valve is closed. This helps mix the

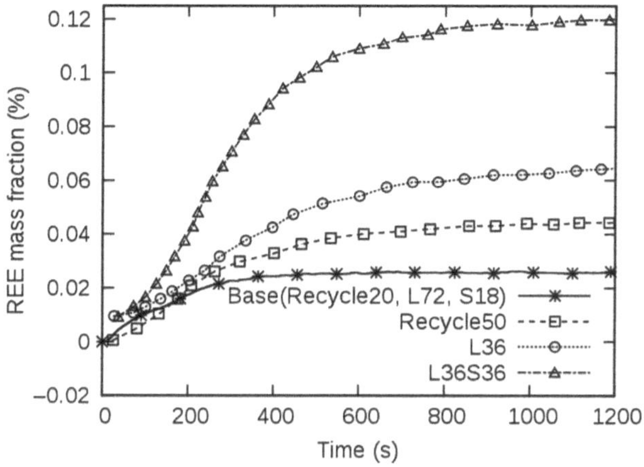

Fig. 41 REE mass fractions in the top outlet liquid under different operating conditions.

particles and is beneficial to REE leaching, which continues even in this bottom section near the outlet.

The REE mass fractions in the top liquid outlet under different operating conditions are compared in Fig. 41. For the base case, 20% of the top outlet liquid stream is recycled into the liquid inlet, and the outlet REE mass fraction reaches a stable profile after about 300 s. However, the outlet REE mass fraction is very small, leading to high energy consumption in the following concentration process. Thus, we modified the operating conditions to increase the final REE mass fraction in the outlet liquid stream. First, the recycled fraction is increased to 50%. Under this condition, the mass fraction of REE is increased to more than 0.04%. According to mass balance, there is ample Mg^{2+} in the liquid; thus, to further increase the REE mass fraction, the liquid solid ratio is reduced by reducing the liquid flow rate and increasing the solid flow rate. Results show that the REE mass fraction can be increased to 0.12%, which is about six times the value of the base case.

4.3.3 Scale-Up of the REE Leaching Reactor

Reactor scale-up is one of the most important tasks for chemical engineers and has been regarded as a major challenge in chemical reaction engineering due to the nonlinear properties of fluid flow, heat transfer, and chemical reactions for single-phase flow reactors. For multiphase flow reactors, however, one should add a little bit of witchcraft (Matsen, 1997 in Knowlton et al., 2005) to make any predictions because these complex reactors seem outside the realm of human understanding. Traditional experiment-based

Table 14 Geometries of the Counter-Current Leaching Reactor at Different Scales

	Small	Medium	Large
Reactor diameter, D_1 (m)	0.08	0.16	0.80
Bottom outlet diameter, D_2 (m)	0.04	0.08	0.40
Height, H_1 (m)	0.04	0.08	0.16
Height, H_2 (m)	0.08	0.16	0.32
Height, H_3 (m)	0.10	0.20	0.40
Height, H_4 (m)	0.48	1.12	2.24
Height, H_5 (m)	0.64	1.28	2.56
Liquid flow rate, L (kg/h)	72	28.8	7200
Solid flow rate, S (kg/h)	18.41	73.64	1841
Solid inventory (kg)	0.409	3.273	163.644
Averaged residence time (s)	80	160	320
Particle diameter (mm)	0.45	0.6	1.2

approaches usually start from cold flow, which focuses on hydrodynamics, and the coupling between mass/heat transfer and hydrodynamics is often neglected. With the help of CFD, reactor scale-up can be accomplished very efficiently by coupling different phenomena, such as reactions, with hydrodynamics. In this part, the small-scale REE leaching reactor studied previously is scaled up in size using CFD–CGHS. The simulation setup is the same as that in Section 4.3.1. The detailed geometries of the reactors at different scales are listed in Table 14.

Fig. 42 shows the snapshots of REE leaching reactors from small to large scales. It graphically compares the scales of the three simulated reactors. In simulation, the reactor is virtually scaled up by 2 and 4 times in the axial direction, and 2 and 10 times in the radial direction. Note that it was not necessary to scale up the reactor height 10 times because the target leaching rate was obtained in a shorter reactor. There is no analytical relationship between reactor size, particle diameter, and performance. For a specified particle diameter, we calculate the residence time in the reactor based on the nonlinear reaction kinetics. Then, based on this residence time, we know the "theoretical" height of the reactor because of a known terminal velocity calculated from the assumed drag law. The actual reactor height

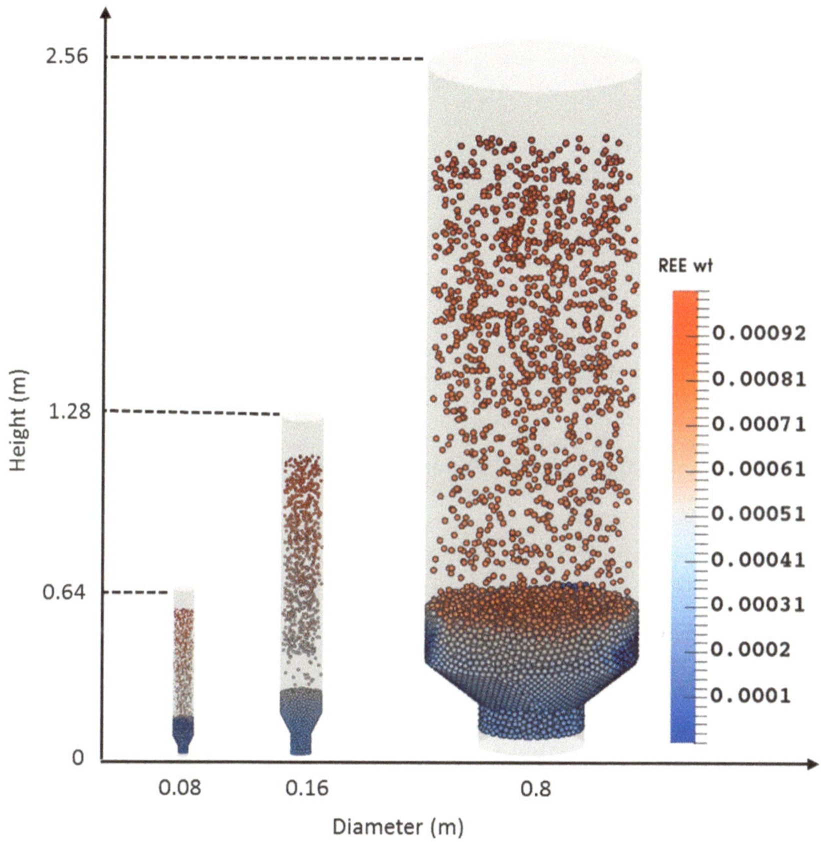

Fig. 42 Snapshots of REE leaching reactor under different scales. *Modified from Lu L, Benyahia S: Thousand-fold speedup of discrete-particle-based computer-aided reactor design and scale-up,* Advanced Materials TechConnect Briefs *(vol 1):363–366, 2017 with permission for the reprinted content. Copyright (2017) Nano Science and Technology Institute.*

is lower because we can allow the particles to stay longer in the dense region of the bottom of the reactor. Postprocessing the simulation results show that the amount of calculated leached REE ions from small to medium to large reactors are 0.36, 1.45, 36.2 kg/day, respectively, based on a targeted 80% extraction efficiency. The distribution of particles shown in Fig. 42 indicates similarities in the particulate dynamics for the three scales of reactors, due to the flow homogeneity that is generally observed in liquid–solids fluidized beds.

From small scale to medium scale, the reactor dimension is doubled in all directions. Thus, the solid and liquid flow rate are increased four times.

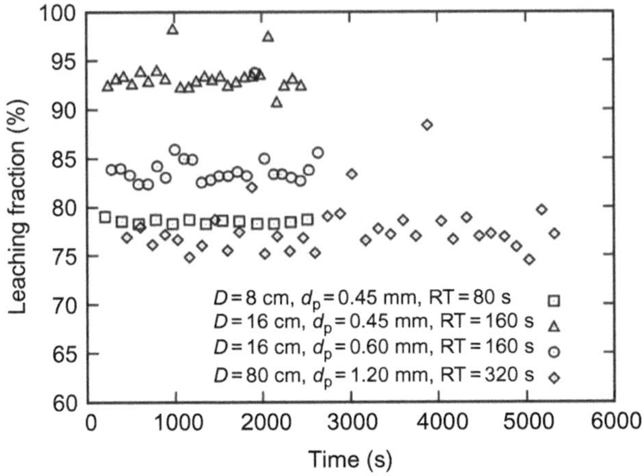

Fig. 43 REE leaching fraction with different scales and particle diameters.

To maintain the same nondimensional dense phase height, the space–time (or averaged residence time) is also doubled. This indicates a large increase in leaching fraction due to the doubled reaction time. As shown in Fig. 43, the leaching rate is increased from about 80% to 93%, which is higher than the targeted leaching fraction of around 80%. Some engineers may be pleased with this result since the performance is better than the target value. However, this outperformance relies on small particle size, which usually requires high energy consumption due to excessive grinding during preparation. Since the leaching reaction rate is directly related to particle size, a larger particle size is also tested. As shown in Fig. 43, for a medium-sized reactor, the particle size can be increased to 0.6 mm and, due to the longer residence time, the leaching fraction is still above 80%. For a large-scale reactor, the particle size is increased to 1.2 mm and the final leaching fraction is about 77%, which is still around the target value. From these simulation results, we can conclude that larger particles can be used for larger reactors. This is highly desirable because using larger particles is more economical due to the energy savings realized by minimized grinding of solid material.

Finally, the REE mass fractions in the top outlet liquid stream under different scales are compared in Fig. 44. A similar REE mass fraction is achieved at the outlet liquid stream at steady-state operation. The only difference among the different scales is the time required to reach this stable state. This is due to the different residence times achieved in the three different reactors. Nevertheless, similar steady operation of these reactors is achieved at the same performance level, even though the particle size increased with

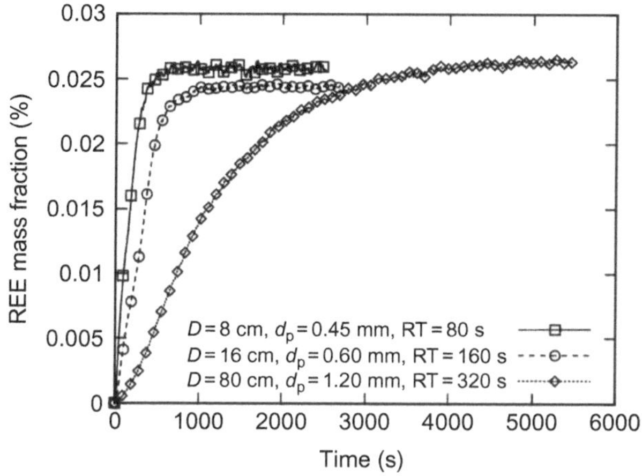

Fig. 44 REE mass fractions in the top outlet liquid stream under different scales.

increasing reactor size. Keeping the same particle diameter for all three reactors would increase the performances of the largest reactor, as previously mentioned.

As with some previous studies described in this chapter, one aspect of these REE simulations is the long simulation time achieved as indicated, for example, in Fig. 44 where thousands of simulation seconds were conducted in the large-scale reactor. The ability to conduct long simulations is something new with discrete particle models and is possible only because of the different assumptions of coarsening detailed in this study. Thus, we finally note in this chapter the ability of the CGHS technique to simulate large-scale reactors for longer time periods under different conditions, leading to scale-up and optimization analysis of multiphase flow reactors used in many industries, as shown in Sections 3 and 4.

5. CONCLUDING REMARKS AND PERSPECTIVES

This chapter provides a detailed description of a coarse discrete particle method with a time-driven hard sphere model developed here for the purpose of industrial-scale multiphase flow reactor simulations. The coarse particle flow dynamics, as well as details of heat transfer with the fluid and other particles/walls, was also provided along with several examples of heterogeneous chemical reactions and mass transfer with the fluid. A brief description of the coupling of this coarse particle method with the fluid and the CFD aspect was also provided in this chapter to give the reader a near-complete picture of the modelling capabilities of the CFD–CGHS method, its development, and usability. The V&V aspect of this method was provided by simulating simplified systems (settling of granular material) and small-scale fluidized bed systems where experimental data and/or numerical data are published and available. The verification step relies heavily on comparison with numerical data obtained by using finer modelling approaches that do not make the coarsening assumptions adopted in this study. Meanwhile, the validation step relies on comparison of our simulation results with available experimental data, most of which are available in small- or laboratory-scale fluidized beds. We extend the validation studies by examining a few industrial-scale reactors where experimental data are also provided in the open literature. We finished this chapter by providing a case where the CFD–CGHS model was used to scale up an REE liquid extraction with process optimization to maximize the REE concentration in the outlet stream. Tackling these large-scale reactors serves a dual purpose of providing a sense of validity of the proposed approach as well as a sense of usefulness and applicability to a wide-range of chemical and energy industries, to name just a few. Thus, by writing this chapter, we hope to increase the usability of CFD methods and help extend their range of validity and also to pinpoint inherent limitations in these methods. It is also our hope that these limitations will encourage more CFD development by academia and industry to ultimately achieve better predictability at the lowest cost possible.

The limitations of CFD and its application to industrial-scale problems are associated with two aspects tackled in this chapter: the complexity of the models and the speed of execution. Extremely complex models are inherently difficult to use and discouraging to apply to practical case studies. Also, slow-to-execute computational models are only practical for small-scale studies and may be too costly to run on today's computers or even supercomputers for industrial-scale problems. For this purpose, we propose the use of simple models with simple to understand assumptions that can produce reasonably accurate results at a speed acceptable to most industrial practitioners. For this purpose, we demonstrated that simulating large-scale reactors can be achieved in just a few days of wall-clock time using minimal computer resources that most research groups possess. Also, because large-scale reactors are often associated with large residence time for particles, the ability to run simulations for thousands of seconds was demonstrated. The speed of the industrial-scale simulations conducted in this study proves the ability of this newly developed technique to tackle not only industrial-scale problems but also their optimization using a rigorous CFD approach instead of traditional chemical process software based on oversimplification of the complex flow dynamics in multiphase flow reactors. This provides the design engineer with an alternative tool to most process optimization software, which are traditionally based on simple heat and mass balances on oversimplified reactor models that usually ignore Newton's second law of motion, i.e., momentum conservation. That said, our intention is not to undermine the usefulness of process design and optimization tools that can simulate an entire plant instead of just a unit operation (fluidized bed) as proposed in this chapter. Given the current computational resources, we are still far from simulating the whole plant with the current CFD approach, therefore increasing simulation speed should be paramount to modelers and simulation tool developers. Increasing the MFIX computation speed to increase its fluid solver speed is a low hanging fruit. A simple simulation of a 3D single-phase jet flow with Re number of 2.78×10^4 proves that the execution speed of the MFIX fluid flow solver is extremely slow. Hence Fig. 45 indicates that the speed of the MFIX fluid solver can be increased significantly by adopting different flow solving algorithms, such as PISO instead of SIMPLE (Barton, 1998; Chow and Cheung, 1997). The PISO algorithm is much faster than SIMPLE because it uses large underrelaxation factors and time steps. Other unpublished findings, the details of which are not provided here for brevity, show the increase in simulation speed due to the linear equation solver, from that used in MFIX to the algebraic multigrid solver used in a different CFD code. With these tweaks to the flow and linear equation solver, it will be possible to achieve much faster results that will appeal even more to industrial users. For example, previous simulation using OpenFOAM coupled with GPU-accelerated coarse-grained CFD–DEM can simulate a pilot-scale CFB riser at almost real-time in 2D (Ge et al., 2015; Lu et al., 2016c). With this speed, the user can make online adjustments to the operating parameters, like superficial gas velocity, and the corresponding hydrodynamics can be observed directly through the display program as shown in Fig. 46.

The most important remaining work not covered in this chapter (to be conducted soon) is a systematic study to determine the accuracy vs speed of the coarsening assumption. Basically, we want to answer the question of how much uncertainty do we introduce in our simulation by assuming a certain parcel size. We should note that this type of uncertainty is related to model assumptions and not input or discretization errors. We simply assume that CFD–DEM results are the most accurate and any deviation is due to model coarsening by adopting parcels instead of real particles. For this

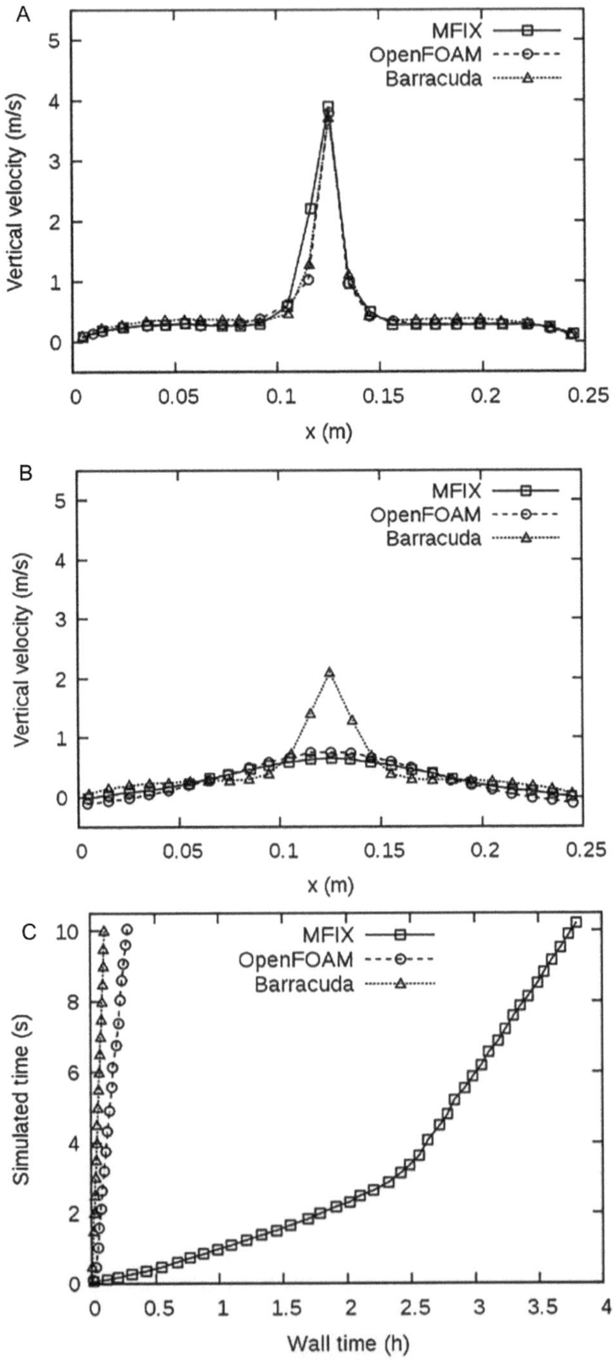

Fig. 45 Single-phase jet flow simulation with a laminar (A) and turbulent (B) models simulated with three CFD codes along with their simulation time (C) conducted on the same single-processor.

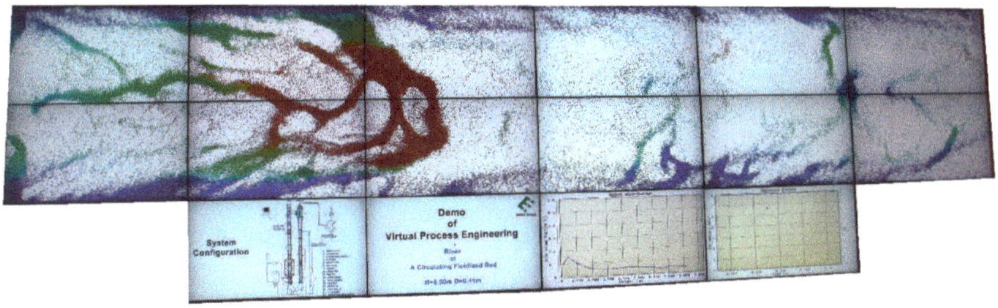

Fig. 46 Interactive simulation of a pilot-scale CFB riser using OpenFOAM coupled with GPU-accelerated coarse-grained CFD–DEM (Ge et al., 2015; Lu et al., 2016c). *Modified from a photograph of the computer virtual experiment at Institute of Process Engineering, Chinese Academy of Sciences.*

purpose, we plan to study the effect of using different parcel sizes on different system geometries and flow rates. The different geometries will explore the effect of system size (or total number of particles) on the coarsening errors. The different flow rates will answer the basics question of quantifying this uncertainty for different flow regimes, such as bubbling, transport, etc. Answering these questions will give the simulation engineer an idea of what level of uncertainty is associated with a certain level of coarsening for his or her application. Apart from what was pursued in this chapter, another application will be pursued in the near future that relates to particles with changing properties, such as PSD, that has many applications in crystallization of a solute as well as polymerization reactors used in the pharmaceutical and chemical industry, respectively. The power of discrete particle methods is that the change of PSD or any other property of the particles is straightforward to code and apply as long as crystallization or polymerization growth rates are known a priori from other studies. Unlike the continuum model or TFM, discrete methods do not require the additional burden of modelling PSD changes through population balance equations. Unlike population balance equations, discrete methods provide continuous size and other property distributions. In addition, if binning the PSD is required for comparison with experimental data, then this will be just a straightforward postprocessing step. As mentioned earlier, the lack of additional model complexity is one of the most attractive aspects of discrete methods. Another important development effort that was not covered in this chapter but should be conducted in the future is the radiative heat transfer. Radiation is important for combustion and high temperature applications for many industries and will be investigated and implemented in the future.

Finally, we encourage readers to download the videos associated with this work because it is difficult to explain all the interesting results that were obtained by simply looking at graphical representation of time-averaged results. We include several animations of our numerical results so that readers have the same experience as we did when looking at and analyzing these results. We also believe that good quality videos (from both simulation and experiments) are very useful educational tools that help students and practitioners understand the often opaque inside of their fluidized bed reactor. Supplementary data to this article can be found online at https://doi.org/10.1016/bs.ache.2017.12.001.

ACKNOWLEDGMENTS

This research was supported in part by an appointment to the National Energy Technology Laboratory Research Participation Program, sponsored by the US Department of Energy and administered by the Oak Ridge Institute for Science and Education. The authors would like to thank the support from Multiphase Flow Science team at National Energy Technology Laboratory. The first author, L.L., would also like to thank Institute of Process Engineering, Chinese Academy of Sciences for supporting his PhD study which has been a major basis for this research.

DISCLAIMER

REFERENCES

Andrews MJ, O'Rourke PJ: The multiphase particle-in-cell (MP-PIC) method for dense particulate flows, *Int J Multiphase Flow* 22:379–402, 1996.

Avidan AA, Shinnar R: Development of catalytic cracking technology. A lesson in chemical reactor design, *Ind Eng Chem Res* 29:931–942, 1990.

Barton IE: Comparison of SIMPLE-and PISO-type algorithms for transient flows, *Int J Numer Methods Fluids* 26:459–483, 1998.

Batchelor GK, Brien RW: Thermal or electrical conduction through a granular material, *Proc R Soc Lond A Math Phys Sci* 355:313, 1977.

Beetstra R, van der Hoef MA, Kuipers JAM: Drag force of intermediate Reynolds number flow past mono- and bidisperse arrays of spheres, *AIChE J* 53:489–501, 2007.

Benyahia S, Galvin JE: Estimation of numerical errors related to some basic assumptions in discrete particle methods, *Ind Eng Chem Res* 49:10588–10605, 2010.

Benyahia S, Syamlal M, O'Brien TJ: Extension of Hill–Koch–Ladd drag correlation over all ranges of Reynolds number and solids volume fraction, *Powder Technol* 162:166–174, 2006.

Benyahia S, Syamlal M, O'Brien TJ: Study of the ability of multiphase continuum models to predict core-annulus flow, *AIChE J* 53:2549–2568, 2007.

Buist KA, Seelen LJH, Deen NG, Padding JT, Kuipers JAM: On an efficient hybrid soft and hard sphere collision integration scheme for DEM, *Chem Eng Sci* 153:363–373, 2016.

Chang J, Wang G, Lan X, Gao J, Zhang K: Computational investigation of a turbulent fluidized-bed FCC regenerator, *Ind Eng Chem Res* 52:4000–4010, 2013.

Chow WK, Cheung YL: Comparison of the algorithms PISO and simpler for solving pressure-velocity linked equations in simulating compartmental fire, *Numer Heat Transf A Appl* 31:87–112, 1997.

De Wilde J, Richards G, Benyahia S: Qualitative numerical study of simultaneous high-G-intensified gas–solids contact, separation and segregation in a bi-disperse rotating fluidized bed in a vortex chamber, *Adv Powder Technol* 27:1453–1463, 2016.

Fogler HS: *Elements of chemical reaction engineering,* New Jersey, 2009, Prentice-Hall International.

Gao J, Lan X, Fan Y, et al: CFD modeling and validation of the turbulent fluidized bed of FCC particles, *AIChE J* 55:1680–1694, 2009.

Garg R, Galvin J, Li T, Pannala S: Documentation of open-source MFIX-DEM software for gas-solids flows, 2012a: From URL, https://mfix.netl.doe.gov/download/mfix/mfix_current_documentation/dem_doc_2012-1.pdf, 2012a.

Garg R, Galvin J, Li T, Pannala S: Open-source MFIX-DEM software for gas–solids flows: part I—verification studies, *Powder Technol* 220:122–137, 2012b.

Garzó V, Dufty JW, Hrenya CM: Enskog theory for polydisperse granular mixtures. I. Navier-Stokes order transport, *Phys Rev E Stat Nonlin Soft Matter Phys* 76:031303, 2007.

Ge W, Li J: Physical mapping of fluidization regimes—the EMMS approach, *Chem Eng Sci* 57:3993–4004, 2002.

Ge W, Lu L, Liu S, Xu J, Chen F, Li J: Multiscale discrete supercomputing—a game changer for process simulation? *Chem Eng Technol* 38:575–584, 2015.

Ge W, Wang L, Xu J, et al: Discrete simulation of granular and particle-fluid flows: from fundamental study to engineering application, *Rev Chem Eng* 33(6):551–623, 2017.

Gidaspow D: *Multiphase flow and fluidization: continuum and kinetic theory descriptions,* 1994, Academic Press.

Grace J, Sun G: Influence of particle size distribution on the performance of fluidized bed reactors, *Can J Chem Eng* 69:1126–1134, 1991.

Grace JR, Taghipour F: Verification and validation of CFD models and dynamic similarity for fluidized beds, *Powder Technol* 139:99–110, 2004.

Gunn DJ: Transfer of heat or mass to particles in fixed and fluidised beds, *Int J Heat Mass Transf* 21:467–476, 1978.

Helland E, Occelli R, Tadrist L: Numerical study of cluster and particle rebound effects in a circulating fluidised bed, *Chem Eng Sci* 60:27–40, 2005.

Hoomans BPB, Kuipers JAM, Briels WJ, van Swaaij WPM: Discrete particle simulation of bubble and slug formation in a two-dimensional gas-fluidised bed: a hard-sphere approach, *Chem Eng Sci* 51:99–118, 1996.

Igci Y, Andrews AT, Sundaresan S, Pannala S, O'Brien T: Filtered two-fluid models for fluidized gas-particle suspensions, *AIChE J* 54:1431–1448, 2008.

Jackson R: *The dynamics of fluidized particles,* 2000, Cambridge University Press.

Jun T, Jingqun Y, Ruan C, Guohua R, Mintao J, Kexian O: Kinetics on leaching rare earth from the weathered crust elution-deposited rare earth ores with ammonium sulfate solution, *Hydrometallurgy* 101:166–170, 2010.

Kirbas G, Kim SW, Bi X, Lim J, Grace JR: In Berruti F, Bi X, Pugsley T, editors: *Radial distribution of local concentration weighted particle velocities in high density circulating fluidized beds, 12th International conference on fluidization*, 2007, pp 71–78.

Knowlton T, Karri S, Issangya A: Scale-up of fluidized-bed hydrodynamics, *Powder Technol* 150:72–77, 2005.

Kunii D, Levenspiel O: *Fluidization engineering,* Newton, MA, 1991, Butterworth-Heinemann.

Kwauk M: *Fluidized leaching and washing,* (in chinese), Beijing, 1979, Science Press.

Kwauk M: *Fluidization–idealized and bubbleless, with applications,* New York, 1992, Ellis Horwood.

Li J, Kwauk M: *Particle-fluid two-phase flow: the energy-minimization multi-scale method,* Beijing, 1994, Metallurgical Industry Press.

Li T, Dietiker J-F, Shahnam M: MFIX simulation of NETL/PSRI challenge problem of circulating fluidized bed, *Chem Eng Sci* 84:746–760, 2012a.

Li T, Garg R, Galvin J, Pannala S: Open-source MFIX-DEM software for gas-solids flows: part II—validation studies, *Powder Technol* 220:138–150, 2012b.

Li T, Dietiker J-F, Shadle L: Comparison of full-loop and riser-only simulations for a pilot-scale circulating fluidized bed riser, *Chem Eng Sci* 120:10–21, 2014.

Lu C, Wang Z: *Fluid catalytic cracking technology,* Beijing, 2002, China Petrochemical Press.

Lu L, Xu J, Ge W, Yue Y, Liu X, Li J: EMMS-based discrete particle method (EMMS–DPM) for simulation of gas–solid flows, *Chem Eng Sci* 120:67–87, 2014.

Lu L, Yoo K, Benyahia S: Coarse-grained-particle method for simulation of liquid–solids reacting flows, *Ind Eng Chem Res* 55:10477–10491, 2016a.

Lu B, Luo H, Li H, et al: Speeding up the modeling of MTO fluidized bed reactor with integration of CRE and CFD, *Chem Eng Sci* 143:341–350, 2016b.

Lu L, Xu J, Ge W, et al: Computer virtual experiment on fluidized beds using a coarse-grained discrete particle method—EMMS-DPM, *Chem Eng Sci* 155:314–337, 2016c.

Lu L, Gopalan B, Benyahia S: Assessment of different discrete particle methods ability to predict gas-particle flow in a small-scale fluidized bed, *Ind Eng Chem Res* 2017(56):7865–7876, 2017a.

Lu L, Konan A, Benyahia S: Influence of grid resolution, parcel size and drag models on bubbling fluidized bed simulation, *Chem Eng J* 326:627–639, 2017b.

Lu L, Morris A, Li T, Benyahia S: Extension of a coarse grained particle method to simulate heat transfer in fluidized beds, *Int J Heat Mass Transf* 111:723–735, 2017c.

Lu L, Li T, Benyahia S: An efficient and reliable predictive method for fluidized bed simulation, *AIChE J* 63:5320–5334, 2017d. https://doi.org/10.1002/aic.15832.

Lu L, Gao X, Li T, Benyahia S: Numerical investigation of the ability of salt tracers to represent the residence time distribution of fluidized catalytic cracking particles, *Ind Eng Chem Res* 56:13642–13653, 2017e.

Merrow EW, Phillips KE, Myers CW: *Understanding cost growth and performance shortfalls in pioneer process plants,* Santa Monica, CA, 1981, Rand Corp.

Moldoveanu GA, Papangelakis VG: An overview of rare-earth recovery by ion-exchange leaching from ion-adsorption clays of various origins, *Mineral Mag* 80:63, 2016.

Ouyang J, Li J: Particle-motion-resolved discrete model for simulating gas-solid fluidization, *Chem Eng Sci* 54:2077–2083, 1999.

Patankar NA, Joseph DD: Modeling and numerical simulation of particulate flows by the Eulerian–Lagrangian approach, *Int J Multiphase Flow* 27:1659–1684, 2001.

Patil AV, Peters EAJF, Kuipers JAM: Comparison of CFD–DEM heat transfer simulations with infrared/visual measurements, *Chem Eng J* 277:388–401, 2015.

Radl S, Sundaresan S: A drag model for filtered Euler–Lagrange simulations of clustered gas–particle suspensions, *Chem Eng Sci* 117:416–425, 2014.

Radl S, Radeke C, Khinast JG, Sundaresan S: In *Parcel-based approach for the simulation of gas-particle flows, 8th International conference on CFD in oil & gas.* Trondheim, 2011, Metallurgical and Process Industries.

Richard P, Nicodemi M, Delannay R, Ribiere P, Bideau D: Slow relaxation and compaction of granular systems, *Nat Mater* 4:121–128, 2005.

Rong D, Horio M: In DEM simulation of char combustion in a fluidized bed, Second International conference on CFD in the minerals and process industries. Melbourne, Australia, 1999, CSIRO, pp 65–70.

Sakai M, Koshizuka S: Large-scale discrete element modeling in pneumatic conveying, *Chem Eng Sci* 64:533–539, 2009.

Sakano M, Yaso T, Nakanishi H: Numerical simulation of two-dimensional fluidized bed using discrete element method with imaginary sphere model, *Jpn J Multiphase Flow* 14:66–73, 2000.

Shaffer F, Gopalan B, Breault RW, et al: High speed imaging of particle flow fields in CFB risers, *Powder Technol* 242:86–99, 2013.

Shi Z, Wang W, Li J: A bubble-based EMMS model for gas–solid bubbling fluidization, *Chem Eng Sci* 66:5541–5555, 2011.

Syamlal M, Benyahia S: High-resolution methods for preserving the sum of mass fractions: improved χ-scheme and an alternative, *Int J Numer Methods Fluids* 73:750–764, 2013.

Syamlal M, O'Brien T: *Derivation of a drag coefficient from velocity-voidage correlation*, Morgantown, West Virginia, 1987, US Department of Energy, Office of Fossil Energy, National Energy Technology Laboratory. April.

Tang Y, Lau YM, Deen NG, Peters EAJF, Kuipers JAM: Direct numerical simulations and experiments of a pseudo-2D gas-fluidized bed, *Chem Eng Sci* 143:166–180, 2016.

Tebianian S, Dubrawski K, Ellis N, et al: Comparison of particle velocity measurement techniques in a fluidized bed operating in the square-nosed slugging flow regime, *Powder Technol* 296:45–52, 2016.

Thakur SC, Ooi JY, Ahmadian H: Scaling of discrete element model parameters for cohesionless and cohesive solid, *Powder Technol* 293:130–137, 2016.

Tian P, Wei Y, Ye M, Liu Z: Methanol to olefins (MTO): from fundamentals to commercialization, *ACS Catal* 5:1922–1938, 2015.

Tsuji Y, Kawaguchi T, Tanaka T: Discrete particle simulation of two-dimensional fluidized bed, *Powder Technol* 77:79–87, 1993.

Wang W, Li J: Simulation of gas–solid two-phase flow by a multi-scale CFD approach—of the EMMS model to the sub-grid level, *Chem Eng Sci* 62:208–231, 2007.

Wang J, Ge W, Li J: Eulerian simulation of heterogeneous gas–solid flows in CFB risers: EMMS-based sub-grid scale model with a revised cluster description, *Chem Eng Sci* 63:1553–1571, 2008.

Wen CY, Yu YH: In *Mechanics of fluidization*, Chemical engineering progress symposium. 1966, pp 100–113.

Xiao Y, Chen Y, Feng Z, et al: Leaching characteristics of ion-adsorption type rare earths ore with magnesium sulfate, *Trans Nonferrous Met Soc Chin* 25:3784–3790, 2015.

Xu BH, Yu AB: Numerical simulation of the gas-solid flow in a fluidized bed by combining discrete particle method with computational fluid dynamics, *Chem Eng Sci* 52:2785–2809, 1997.

Xu J, Qi HB, Fang XJ, et al: Quasi-real-time simulation of rotating drum using discrete element method with parallel GPU computing, *Particuology* 9:446–450, 2011.

Yang N, Wang W, Ge W, Li J: CFD simulation of concurrent-up gas–solid flow in circulating fluidized beds with structure-dependent drag coefficient, *Chem Eng J* 96:71–80, 2003.

Zhou ZY, Yu AB, Zulli P: Particle scale study of heat transfer in packed and bubbling fluidized beds, *AIChE J* 55:868–884, 2009.

Zhou ZY, Yu AB, Zulli P: A new computational method for studying heat transfer in fluid bed reactors, *Powder Technol* 197:102–110, 2010.

CHAPTER THREE

Spatial Filtering for Scale Bridging and Its Application to Transport in Dense Particle Beds

Stefan Radl[1], Federico Municchi
Institute of Process and Particle Engineering, Graz University of Technology, Graz, Austria
[1]Corresponding author: e-mail address: radl@tugraz.at

Contents

Advances in Chemical Engineering, Volume 53
ISSN 0065-2377
https://doi.org/10.1016/bs.ache.2018.03.001

Abstract

The method of spatial filtering as a universal approach to support closure development for models of dense fluid–particle systems is presented. Starting from a set of governing equations, an overview of closures that are required at different levels of a model hierarchy is provided. A focus is on closures to be used in mesoscale models, highlighting recent developments that aim at the description of the variability of exchange coefficients, as well as wall effects. Furthermore, attempts to describe the anisotropy of mesoscale stresses, as well as the effect of cohesive particle–particle interactions in the context of macroscale models are summarized. Computational aspects, together with selected verification cases are documented to stimulate a broader use of the filtering approach. Finally, three future research directions that appear most fruitful are outlined. These final thoughts may help to establish more reliable and robust multiphase flow models for reactive systems.

1. INTRODUCTION AND TRENDS

1.1 Filtering as Strategy for Scale Bridging

Efforts to model the performance of chemical reactors have taken a turn once detailed flow models became available: nowadays, modelling and simulation of flow plays a central role in the development of equipment in the chemical industry, being an integral part of many roadmaps and perspectives (2017). This is especially true for dense fluid–particle flows that are in the focus of this chapter. Such multiphase flows are central to many industrial applications such as (i) the wide range of processes that rely on fluidization, (ii) packed bed reactors, (iii) solids handling and conveying, or (iv) particle fractionation and separation to name just a few examples.

Certainly, the influential perspective of Sundaresan was an eye-opener for many researchers and resulted in a number of breakthrough papers in the field of chemical engineering and fluid mechanics (Sundaresan, 2000). Key statements in this perspective—e.g., related to the fact that "…work is under way in several research groups to include the effect of particle-size distribution…"—are still valid today. It stresses the fact that differences in particle size—on the first view a trivial twist in the parameters that describe a gas–particle system—add significant complexity when attempting to model transport phenomena in dense particulate systems. This is an addition to complex flow structures that arise spontaneously due the inherent instability of the fluidized state (Fullmer and Hrenya, 2017): an ensemble of homogeneously dispersed particles will spontaneously form clusters when allowed to sediment freely. Thus, one has to generally except that particles in relevant flow situations have a spatially nonuniform distribution (this is even the case in packed beds!). This is the exact reason why spatial averaging over a region with a certain size (or "filtering") plays such an important role for these multiphase flows: in order to arrive at a meaningful description, one must deeply appreciate the difference between local and average quantities.

Since the mathematics behind spatial averaging has been summarized quite extensively (Whitaker, 1999), it appears that little needs to be said about how to apply filtering in the field of chemical engineering. Unfortunately, the opposite is true: filtering has attracted renewed interest in the wake of a specific type of flow simulations: large-scale Euler–Lagrange (EL) simulations of dense gas–particle flows became increasingly fashionable over the past 5 years (Capecelatro and Desjardins, 2013; Radl and Sundaresan, 2014). Adopting the EL strategy over the commonly used Euler–Euler approach (EE; often also called "two-fluid model,") appears to be a trivial modification. But this is not the case, as can be shown by two important facts:

(i) EL models necessitate closures on a per-particle level, while EE models are based on closures that describe ensemble-average quantities and

(ii) EL models enable simulations of cohesive systems, and systems in which intraparticle transport processes play a central role.

Next, a brief description of the hierarchy of models is introduced in Section 1.2. Closures derived by filtering can be classified within this hierarchy. One could call models that are based on such a hierarchy "multiscale models" (van der Hoef et al., 2008). Since the exact definition of "multiscale" differs between different authors, and is largely irrelevant for the present contribution, this wording is dropped in what follows—it is enough to accept that filtering is a universal strategy that can be used to connect models at various levels of the hierarchy introduced in Section 1.2. In Section 1.3, an outline of the computational techniques often employed in the solution of these models is provided.

Data science challenges arising from filtering operations are briefly discussed in Section 1.4.

The remaining of this chapter is structured as follows: Section 2 summarizes models, as well as the most relevant types of closures laws that are employed to describe dense gas–particle suspensions. Section 3 discusses some algorithmic aspects with a focus on postprocessing tools for spatial averaging. Section 4 is dedicated to ideas on how to verify the implementation of such algorithms. In Sections 5 and 6 recent developments in the field of mesoscale models are summarized. Specifically, a reflection on newly proposed closure laws that are required as input to these models, as well as closures that can be developed on the basis of such models is provided. Section 7 completes this chapter with a conclusion and an outlook.

1.2 Hierarchy of Models and Their Connection

1.2.1 Chemistry and Material Models

An overview of models used in typical simulations of reactive gas–particle suspension flows is provided in Fig. 1. First, an essential classification must be made, specifically in (i) chemistry and materials models, as well as (ii) multiphase flow models. For the description of models in category (i) a unified nomenclature was recently developed (de Baas and Rosso, 2015). These models

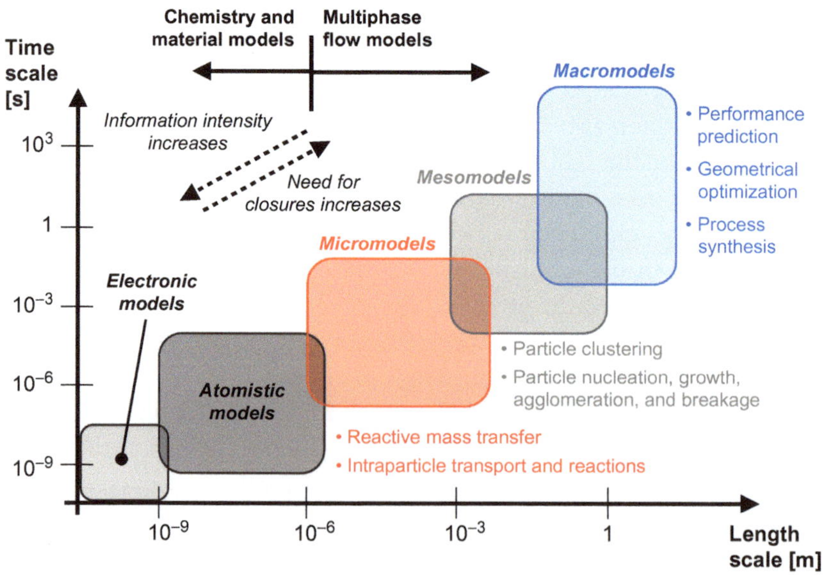

Fig. 1 Overview of models used in typical reactive fluid–particle flow applications categorized by a typical time and length scale.

describe phenomena occurring on a subatom to a super-molecular scale, and hence are predominantly used by chemists and physicists. Chemistry and material models can be subcategorized into (a) electronic, (b) atomistic, (c) mesoscopic,[a] as well as (d) continuum models (for clarity, subcategories c and d are not indicated in Fig. 1). A classical application of electronic and atomistic models is the development of reaction mechanisms and associated kinetics. Thus, the result of simulations using these models is an understanding of how chemical reactions proceed, which typically includes reaction stoichiometry, reaction enthalpies, and rate laws. Another application is the direct simulation of diffusion via atomistic models, e.g., tracking the motion of a molecule in a matrix consisting of other molecules.

1.2.2 Flow Models

Multiphase flow models are built on top of materials models, and use the *output* of chemistry and material property models to close the governing equations. An overview of relevant applications of flow models was provided recently by Pannala et al. (2011).

Multiphase flow models typically follow a less strict nomenclature as that introduced for chemistry and material models (de Baas and Rosso, 2015). In most cases, flow models are categorized based on the size of the phenomena they describe (Padding, 2013). For example, the NanoSim project (2017) adopted a nomenclature that relied on the name of a typical item (e.g., a particle) to define a model (see Fig. 2).

Even simpler definitions are widely used in the context of fluid–particle systems, often leading to confusion. For example, the process model categories introduced by Padding (2013) are based on the assumption that atoms are the smallest entities. However, when describing suspension flows in engineering applications, it would be more natural to consider particles as being the smallest entities. When following such an engineering concept, the following categories of models are typically used, a wording which is also adopted in the present work:

(1) "*Micro*"(*scale*) *models* are concerned with phenomena that happen on a subparticle or particle scale. The publications of Khinast and coworkers (Lekhal et al., 2001a, b) dealing with drying of particles (i.e., intraparticle transport of heat and mass) are a prime example of such models. Another example in the same category is the large body of direct numerical

[a] Not to be confused with the wording "mesoscale models" (Fox, 2012), a wording that is adapted in what follows for flow models.

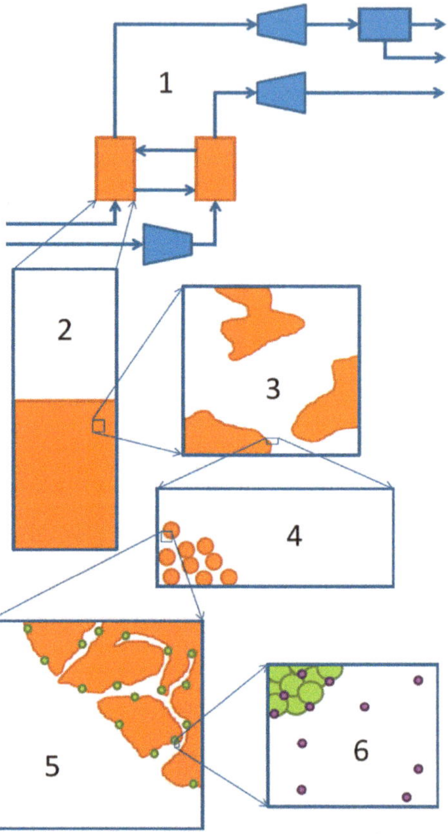

Fig. 2 An example for the definition of a multiscale model in the context of process models (this definition was developed during the NanoSim project (2017) and distinguishes six scales: (1) system scale, (2) equipment scale, (3) cluster scale, (4) particle scale, (5) intraparticle pore scale, and (6) atomistic scale. *Figure adapted from NanoSim—a multiscale simulation-based design platform for cost-effective co2 capture processes using nano-structured materials (Annex I—Description of work). Trondheim, Norway, 2013.*

simulation (DNS) models that have been used to predict fluid–particle transfer rates of momentum, heat, and mass (Deen and Kuipers, 2014a; Feng and Michaelides, 2009; Municchi and Radl, 2017; Sun et al., 2015). Most "microscale" models do not allow particles to move (Holloway et al., 2012; Lekhal et al., 2001a), e.g., transport phenomena within a particle are separated from the particle's convective transport. However, some studies assign a specific speed to individual particles, but freeze their relative position (Holloway et al., 2010), or allow particles to move freely in order to capture effects due to either (i) the perturbed microstructure or (ii) the inhomogeneous distribution of the particles in the computational domain (Derksen, 2014b; Derksen and Sundaresan, 2007; Uhlmann and Doychev, 2014).

(2) "*Meso*"(*scale*) *models* describe phenomena on intermediate scales. In the context of particulate systems this includes phenomena that are governed by the behavior of an *ensemble* of particles (Fox, 2012, 2017). A well-known example of a mesoscale model is the celebrated Boltzmann equation to describe the evolution of a particle cloud in space, time, and velocity coordinates. For fluid–particle systems involving particles that are allowed to change their properties—also referred to as particle-internal coordinates, e.g., size, density—the so-called generalized population balance equation is the relevant mesoscale model for the particle phase (Fox, 2012). The core variable that describes this mesoscale model is the number density function (NDF). This model can be solved by a variety of solution strategies, e.g., a moment-based (Eulerian) method or a Lagrangian method that approximates the NDF with a finite number of computational parcels (i.e., "packages of particles"). The latter strategy typically requires tracking a large number of particles or parcels. An idea is to exploit Graphic Processing Units (GPUs) to reduce the computation time when using a Lagrangian method (Radeke et al., 2010; Radl et al., 2011) to evaluate a mesoscale model.

(3) "*Macro*"(*scale*) *models* aim on describing phenomena that occur on an equipment scale. Often, such models may even describe a whole system (e.g., a connection of mixers, splitter, reactors, and separators). Thus, they aim on picturing effects that originate from the connectivity of multiple pieces of process equipment. While the governing equations that define such macroscale flow models are meanwhile established (Schneiderbauer, 2017), fundamental questions still remain how to formulate physically realistic closures for these equations. This is especially true for reactive systems in which closures for average heat (Lane et al., 2016) and species transport, as well as the consumption/production rate of chemically reacting species must be considered (Agrawal et al., 2013; Holloway and Sundaresan, 2012). The adaptation of these closures in models is slow, and only recently first steps have been made to build rigorous macromodels for reactive systems (Zhu et al., 2016). Macroscale flow models for polydisperse systems are equally challenging, and only the subgroup of bidisperse systems has been analyzed so far in greater detail (Holloway and Sundaresan, 2014; Holloway et al., 2011).

1.2.3 Connection of Models

There is a key need for the connection of individual simulation tools, a fact that is now widely accepted by different research communities (Rantanen and Khinast, 2015). Such connection tools should mainly stimulate the convergence of disciplines (e.g., pharmaceutical science and engineering), and hence their main task is to provide a single "language" for information and knowledge exchange. Simulation tools—together with such "connectors"—then define a unified simulation framework with the vision to concertize efforts of researchers that use different tools. A popular example of such a framework is KNIME (Berthold et al., 2008), which uses a graphical interface to document workflows that connect a large variety of simulators and tools. Also, the framework tool "Porto" (Hagelien et al., 2015) has followed similar ideas. Most relevant for the present text is the tool CPPPO (Municchi et al., 2016), which realizes highly efficient spatial averaging of field data to support researchers during closure development. Next, an attempt to summarize computational solution strategies for solving the underlying model equations is made. This is done to illustrate the sources of data which are available to connection tools.

1.3 Computational Strategies for Solving the Model Equations

Based on a well-established set of equations that describe the motion of fluids and particles (see Section 2), a large array of computational strategies has been developed for their solution. The key challenge when evaluating the model equations is to consider the interaction of a large number of discrete entities (i.e., the particles) with a fluid (and vice versa), as well as with themselves. This requires thoughts on how to solve a large system of equations efficiently. Traditionally, an Euler–Euler (EE)-based description was followed (Anderson and Jackson, 1967), and this type of models is still under active development. However, a general trend toward an Euler–Lagrange (EL)-based description is observable. This is certainly motivated by the need to account for variable particle properties that are of central importance for chemical engineering applications, as well as polydisperse flows in a much wider context.

Table 1 provides a summary of the most significant developments in computational strategies to solve EE and EL models. A clear focus is thereby on Euler–Lagrange methods. It is important to note that these computational approaches can be used at different levels of the model hierarchy detailed in Section 1.2.

Table 1 Overview of Computational Approaches for Simulating Fluid–Particle Flows

Approach and Representative Publication(s)	Particle (or Parcel) Collision Tracking	Enduring Contact Treatment	(Short) Collision Treatment	Particle–Turbulence Interaction	Advantages and Drawbacks
1. Euler–Euler approach (Anderson and Jackson, 1967; Desjardins et al., 2008; Ding and Gidaspow, 1990; Fox, 2008; Iddir and Arastoopour, 2005; Jackson, 2000; Kong and Fox, 2017; Kong et al., 2017; Passalacqua and Fox, 2011; Schneiderbauer et al., 2012b; Yuan and Fox, 2011)	—	Mature "frictional" models, no rolling, cohesion in principle possible	Yes, KTGF-based	Yes	+ widely applicable + mature numerical models − treatment of particle-based properties (size, composition) adds significant complexity
2. Euler–Lagrange: Parcel-based, no parcel collision tracking (e.g., MP-PIC, DDPM) (Cloete et al., 2012; Popoff and Braun, 2007; Rourke and Snider, 2014; Snider, 2001)	No	Heuristic particle pressure terms (friction is challenging), no rolling, no cohesion	Yes, Heuristic or KTGF-based	No	+ adding particle-based properties is easy − poor performance for extremely dense flows
3. Euler–Lagrange: Parcel-based with statistical parcel collision tracking (Sommerfeld, 2003)	Yes, but only statistical	No	Yes, using "virtual particles"	Yes, but only fluid-to-particle	+ adding particle-based properties is easy − poor performance for moderately dense flows
4. Euler–Lagrange: Parcel-based with parcel collision tracking (e.g., "DPM") (Bierwisch et al., 2009; Patankar and Joseph, 2001; Radl et al., 2011; Sakai and Koshizuka, 2009)	Yes	Yes, rolling and cohesion in principle possible	Yes, but poor performance	No	+ adding particle-based properties is easy − parcel collisions in dilute flows are challenging

Continued

Table 1 Overview of Computational Approaches for Simulating Fluid–Particle Flows—cont'd

Approach and Representative Publication(s)	Particle (or Parcel) Collision Tracking	Enduring Contact Treatment	(Short) Collision Treatment	Particle–Turbulence Interaction	Advantages and Drawbacks
5. Euler–Lagrange: Particle collision tracking, fluid flow is not resolved (Capecelatro and Desjardins, 2013; Capecelatro et al., 2014; Girardi et al., 2016; Goniva et al., 2012; Jajcevic et al., 2013; Kloss et al., 2012; Oschmann et al., 2016; Radl and Sundaresan, 2014; Wu et al., 2017)	Yes	Yes, rolling and cohesion models	Yes, direct	Yes	+ widely applicable + direct consideration of particle-based properties (intraparticle profiles can be considered) − closures at the particle level required, which are not well developed − high computational cost
6. Euler–Lagrange: Particle collision tracking, fluid flow is resolved (Deen and Kuipers, 2014a, b; Derksen and Sundaresan, 2007; Schmidt and Nikrityuk, 2012)	Yes	Yes, rolling and cohesion models	Yes, direct	Yes	+ direct calculation of fluid–particle transfer (momentum, heat, mass, etc.) + appealing for nonspherical particles + direct consideration of particle-based properties (intraparticle profiles can be considered) − extremely high computational cost

7. Lagrangian: Parcel-based using per-parcel stress description ("Stress-based DEM", SDEM) (Egholm, 2007; Egholm et al., 2007)	Yes	Yes, via constitutive laws	No	No	+ dense flows + adding particle-based properties is easy − currently no second phase
8. Lagrangian: Parcel-based using Smoothed Particle Hydrodynamics (Xiong et al., 2011; Yuu and Umekage, 2008)	No	Constitutive law	No	No	+ dense flows + adding particle-based properties is easy − collisions in dilute flows are challenging
9. Euler–Euler/Euler–Lagrange Hybrid (Pirker et al., 2009; Schellander et al., 2013; Schneiderbauer et al., 2016)	No (only test particles)	Same as Euler–Euler approach	Similar to Euler–Euler approach	Yes	+ lower computational cost compared to Euler–Lagrange based approach + adding particle-based properties is easy − complex implementation − limited applicability so far

1.4 Aspects From Data Science

Filtering is connected to the manipulation of a large amount of data, and hence it is instructive to consider aspects from data science. Beck et al. (2016) provide an interesting perspective on how to exploit tools from the field of data science in chemical engineering applications. In the context of particle technology applications, the main drivers for such thoughts are the latest developments in the field of (i) particle flow simulators, as well as (ii) imaging techniques (e.g., optical, MRI, or micro CT). These developments lead to an explosion of the available data volume. For example, considering the increasing availability of computational resources, the amount of data generated by simulators (in case dumped to disk in its full extend) would—by far—surpass current resources for data storage. This necessitates deeper thoughts on how to efficiently filter (both temporarily and spatially) data, a need that goes beyond calculating simple running averages.

An example of such advanced data analysis, which is of key relevance in the context of the present contribution, is the work of Igci et al. (2008). This work highlighted for the first time the importance of sophisticated data analysis, namely, the consideration of the effect of filter size on spatially averaged statistics. These ideas have also been followed by other groups (Ozel et al., 2013; Parmentier et al., 2011), establishing a new level of physical insight with respect to phenomena that need to be modeled. Considering the sheer amount of data needed to build a comprehensive closure, for example, the effective drag force experienced by a clustered suspension (Radl and Sundaresan, 2014), such data analysis is preferably performed *during* a simulation run. This need is also illustrated by more recent simulations which incorporate more than 30×10^6 particles in a single simulation (Capecelatro et al., 2014; Ozel et al., 2016).

An additional thought is related to the interpretation of the large amount of data collected with, for example, detailed suspension flow simulations. Clearly, there is a need to formalize the workflow for such an interpretation in order to automate it to a certain degree. Literature indicates that in other areas such "workflow tools" already exist: for example, KNIME (Berthold et al., 2008) is extensively used in the pharmaceutical industry to define workflows for, e.g., data reading, analysis, and report generation. Similarly, machine learning tools, e.g., Tensorflow (Abadi et al., 2015), are used with increasing frequency to extract knowledge from data.

2. MATHEMATICAL FORMULATION

2.1 Euler–Euler Models

Euler–Euler (EE) models describe both the fluid and the particulate phase with transport equations on a globally fixed coordinate system—particles are not tracked in space and time. Instead, the distribution of particle phase properties (e.g., speed, size, composition, etc.) is considered.

The accurate solution of Euler–Euler models for polydisperse suspension flow (Marchisio and Fox, 2013) is still a subject of mathematical research. The reason is that including internal particle coordinates (e.g., the size, density, or moisture content) adds one (or more) dimensions to the governing equations. This leads to a mathematical challenge that is typically not solved directly, but rather simplified first and then solved (e.g., using moment-based methods; Fox, 2017).

A much more straightforward route (in the context of EE models) can be taken in case one accepts the validity of a mono-sized particle population, where only average particle speeds and the granular temperature are used to describe the true particle velocity distribution (Ding and Gidaspow, 1990; Gidaspow, 1994). In this case, the governing equations that define such EE models (Anderson and Jackson, 1967) are comparably easy to solve. Despite those assumptions, EE models are already able to model mesoscale phenomena, e.g., clustering in fluidized beds.

From the perspective of a chemical engineer interested in reacting suspensions, EE models currently impose a key limitation: intraparticle transport phenomena need to be approximated (Stark et al., 2016; Wurzenberger et al., 2002). This is less attractive for rigorous simulations of transport in a variety of reactive systems that are dominated by such intraparticle effects.

2.2 Euler–Lagrange Models

EL models are based on the idea that for each phase the most natural description is chosen: particle motion is described by a Lagrangian approach that tracks each entity individually (either a particle or a parcel), and fluid motion (i.e., the motion of the continuous phase) is described by an Eulerian approach.

For suspensions relevant in this chapter (featuring particle volume fractions up to the close packing limit), the DEM (or "soft-sphere" approach) is employed to track particles and evaluate their interaction. This is motivated by the ability to model enduring particle–particle collisions. For modelling the interaction of the particles with the fluid phase, two modelling approaches exist: (i) particle-resolved EL models and (ii) particle-unresolved EL models.

2.2.1 Particle-Resolved EL Models

These models are EL-based and designed to predict the details of the flow near the fluid–particle interface. These models are discussed in greater detail in Sections 2.3.1 and 2.4.1. These "particle-resolved" (PR) models necessitate significant computational resources to predict the flow since the domain occupied by the fluid phase is discretized into control volumes that are much smaller than the particles (see left panel of Fig. 3). The use of PR-EL models is justified in case one of the following ideas is followed:

(i) In case one aims on deriving closure models for, for example, forces acting on individual particles (Beetstra et al., 2007; Municchi and Radl, 2017).

(ii) In case no suitable closure model (e.g., to predict the force acting on a particle) for the particle-unresolved (PU) approach (see Section 2.2.2) exists. This is typically the case when nonspherical and/or deformable particles are involved, or the flow is confined by walls. Examples for the latter include the flow in microchannels (Di Carlo et al., 2009).

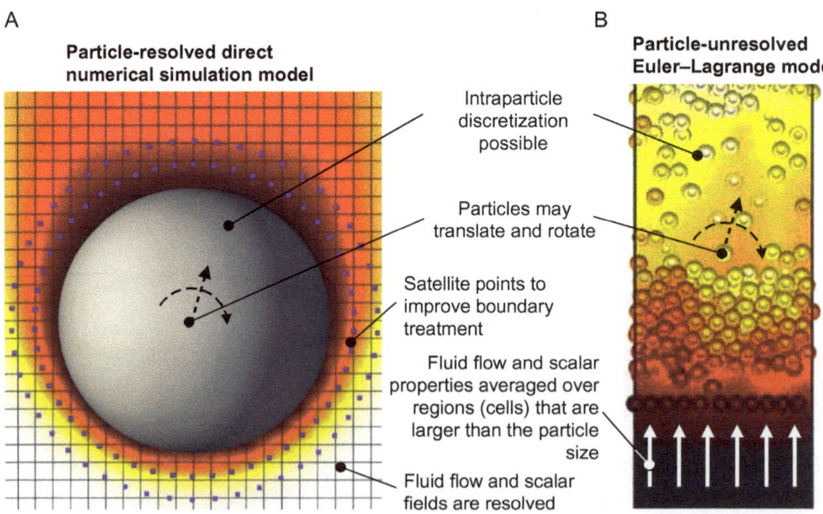

Fig. 3 Idea behind a "particle-resolved" (PR)-EL model that aims on predicting flow details in the void space of a suspension (*left panel*), as well as concept of a "particle-unresolved" (PU)-EL model (*right panel*; *colors* indicate the distribution of a transported scalar quantity, e.g., the temperature).

(iii) In case the details of fluid flow significantly affect the behavior of the suspension, and a PU-EL approach would yield unreliable results. This is typically the case for suspensions in which the fluid-to-particle density ratio is in the order of unity or larger. Examples are liquid–particle suspensions, emulsions, or bubbly flows.

(iv) In case one is interested in the dispersion of small particles in a strongly polydisperse system. This is because small particles will mainly follow the flow around larger particles, necessitating a detailed description of fluid flow in the void space. An example is the modelling of filtration processes (Kolakaluri, 2013; Kolakaluri et al., 2015). Similarly, in case one is interested in how quickly a scalar quantity (e.g., a dissolved substance) is dispersed in the continuous phase of a suspension (Derksen, 2008b), one will favor a PR-EL approach.

2.2.2 Particle-Unresolved EL Models

These models are a simplification, where only spatially averaged transport equations for the continuous phase are solved (see right panel of Fig. 3, as well as Fig. 4). While drastically reducing the demand of computational resources to solve the governing equations, PU-EL-based models have lower predictive capabilities. Fortunately, for most systems the adoption of a PU-EL approach is still appropriate. For example, in gas–particle flows the inertia of the continuous phase (i.e., the gas) is negligible compared to that of the particles (an example is a recent work that benchmarked data

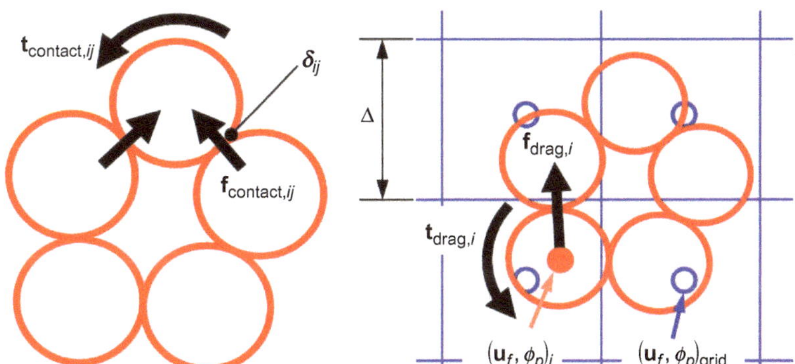

Fig. 4 Idea behind a PU-EL model using the DEM as the Lagrangian submodel (*left panel*: illustration of contact force and torque calculation; *right panel*: illustration of the computational Eulerian grid in *blue*, as well as particle-based force and torque calculation; t is a torque, f is a force, δ is an overlap, u is a velocity, Δ is a distance, and ϕ is a volume fraction).

originating from PU-EL and PR-EL for dilute suspensions; Horwitz et al., 2016). Hence, small-scale fluid velocity fluctuations (that are not directly resolved in PU-EL, but may be modeled) do not critically affect the overall system behavior. Finally, it should be mentioned that often a PU-EL is chosen, simply due to computational limitations, and limited predictive capabilities are accepted.

2.3 Governing Equations for Momentum Transport

The governing equations are derived from basic balance equations of mass, momentum, and a scalar quantity (for a rigorous discussion see the Appendix: note that the indication of average quantities is different in this appendix since two levels of spatial averaging are performed in what follows). In particular, in the following equations for the momentum transport in the fluid and particulate phase are summarized, together with the equations for mass conservation in an incompressible fluid. Transport equations of scalar quantities are discussed in Section 2.4.

2.3.1 Continuous Phase: Particle-Resolved Approach

Adopting the PR-EL approach the original form of the Navier–Stokes equations for the description of the fluid motion can be used. In the following description, the fluid is assumed to be incompressible having a constant dynamic viscosity μ. The transport equations for mass conservation and the acceleration of the fluid can then be formulated as:

$$\nabla \cdot \mathbf{u}^{\circ} = 0 \tag{1}$$

$$\rho_f[\partial_t \mathbf{u}^{\circ} + \nabla \cdot (\mathbf{u}^{\circ}\mathbf{u}^{\circ})] = -\nabla p^{\circ} + \mu \nabla^2 \mathbf{u}^{\circ} + \rho_f \mathbf{g} + \mathbf{F}^{\Omega} \tag{2}$$

Here, ρ, μ, $\mathbf{g}$, p, and $\mathbf{u}$ are the fluid density, dynamic viscosity, gravity, pressure, as well as the fluid velocity, respectively. $\mathbf{F}^{\Omega}$ is a volumetric boundary forcing term that is used in some formulations, e.g., when adopting the immersed boundary method (IBM) to solve this transport equation (Municchi and Radl, 2017). The superscript circle in Eqs. (1) and (2) indicates point variables (i.e., they are not averaged, but local in the sense of the continuum hypothesis), to distinguish them from their local mean counterparts introduced in Section 2.3.2.

In case Eqs. (1) and (2) are solved directly (i.e., all details of the spatiotemporal evolution of the flow are resolved), the approach is called PR-direct numerical simulation (PR-DNS). For cases of particle-resolved simulations characterized by very large Reynolds numbers, turbulence models to close the stress in the momentum transport equation might be adopted (Dixon, 2017). Despite being resolved on the particle level, those cannot be called DNS, since not all details of the flow are resolved.

2.3.2 Continuous Phase: Particle-Unresolved Approach

The following set of governing equations is obtained by replacing point variables as introduced in the PR-EL approach (i.e., the variables used in Section 2.3.1) by local mean variables obtained from an averaging operation. A complete description of this averaging operation can be found in standard texts (Anderson and Jackson, 1967) and is illustrated with equations in Appendix. It is to highlight, that this averaging operation, while similar in its concept, is different from the spatial filtering operation discussed in Section 2.3.5. Thus, local mean variables and filtered quantities shall not be confused. Again, as for the PR-EL approach, the fluid is assumed to be incompressible having a constant dynamic viscosity μ. The resulting set of equations for local mean flow quantities reads:

$$\partial_t \phi_f + \nabla \cdot \left(\phi_f \mathbf{u} \right) = 0, \tag{3}$$

$$\rho_f \left[\partial_t \left(\phi_f \mathbf{u} \right) + \nabla \cdot \left(\phi_f \mathbf{u}\mathbf{u} \right) \right] = -\nabla p + \nabla \cdot \boldsymbol{\tau}_f + \varphi_f \rho_f \mathbf{g} + \boldsymbol{\Phi}_{p \to f}. \tag{4}$$

Here, ϕ is a volume fraction, and $\boldsymbol{\tau}$ is a stress tensor. Subscript f denotes the fluid phase, and subscript p denotes the particulate phase. The total particle–fluid interaction force per unit volume of the suspension, i.e., $\boldsymbol{\Phi}_{p \to f}$, can be decomposed into a number of force components. The most natural and often adopted choice of decomposition for fluid–particle systems is (i) a generalized buoyancy term, as well as (ii) a drag term. Introducing this decomposition can result to a formulation differing from the above momentum transport equation (Zhou et al., 2010), which is, however, equivalent to Eq. (4). This decomposition is discussed in Section 2.5.1.

The volumetric total particle–fluid interaction force $\boldsymbol{\Phi}_{p \to f}$ must be modeled due to the description of momentum transport via local mean variables. Typically, an approach that uses a weighted sum of particle–fluid forces $\mathbf{f}_{p \to f}$ acting on a set of particles within a computational cell is adopted. Specifically, one may use:

$$\boldsymbol{\Phi}_{p \to f, j} = \frac{1}{V_{\text{cell}, j}} \sum_{i \in V_{\text{cell}, j}} w_{ij} \mathbf{f}_{p \to f, i}. \tag{5}$$

Here, $V_{\text{cell},j}$ is the volume of the computational cell j, i is the particle index, and w_{ij} is a weighting function.

2.3.3 Dispersed Phase: Eulerian Description

The fidelity of models for dispersed-phase momentum transport has recently undergone a number of improvements, where the most commonly used formulations are summarized below. The interested reader is referred to the work of Schneiderbauer et al. or Kong and coworkers for a more rigorous formulation (Kong and Fox, 2017; Kong et al., 2017; Schneiderbauer et al., 2012b).

$$\partial_t \varphi_p + \nabla \cdot \left(\varphi_p \mathbf{v}\right) = 0 \tag{6}$$

$$\rho_p \left[\partial_t \left(\phi_p \mathbf{v}\right) + \nabla \cdot \left(\phi_p \mathbf{v}\mathbf{v}\right)\right] = -\varphi_p \nabla p + \nabla \cdot \boldsymbol{\tau}_p + \varphi_p \rho_p \mathbf{g} - \boldsymbol{\Phi}_d \tag{7}$$

Here, $\mathbf{v}$ is the particle speed, which in this case is a local quantity in the sense of the continuum hypothesis. Note, that in the above equation only the drag component $\boldsymbol{\Phi}_d$ of the total particle–fluid interaction force $\boldsymbol{\Phi}_{p\to f}$ was used. The remaining buoyancy contribution is shown as the first term on the right hand side of Eq. (7). The particle-phase stress tensor $\boldsymbol{\tau}_p$ used in the above equation must be closed. This is typically done by decomposing it into a kinetic theory-based and frictional contribution (Schneiderbauer et al., 2012b).

An important and often neglected detail when solving the above momentum transport equation is the boundary condition imposed at a wall. The study of Cloete et al. (2016) summarizes recent progress in this field. Specifically, this study suggests that the more rigorous closures for wall treatment derived by Schneiderbauer et al. (2012a) should be used. This removes the need for adjusting the specularity coefficient in the more commonly applied (but less rigorous) set of boundary conditions (Johnson and Jackson, 1987).

2.3.4 Dispersed Phase: Lagrangian Description

2.3.4.1 Translational Motion

Assuming a constant particle mass m, the governing equation for particle motion, i.e., Newton's equations of motion, reduces to the following ordinary differential equation for each particle i:

$$m_i d_t \mathbf{v}_i = \sum_j \mathbf{f}_{j,i} \tag{8}$$

The remaining challenge is to close all forces $\mathbf{f}_{j,i}$ that originate from different physical phenomena j. In parallel to the above evolution equation for the particle speed, the particle position needs to be updated:

$$d_t \mathbf{x}_i = \mathbf{v}_i \tag{9}$$

2.3.4.2 Rotational Motion

The balance of angular momentum requires that (in a global coordinate system):

$$d_t(\mathbf{I}_i \cdot \boldsymbol{\omega}_i) = \sum_j \mathbf{t}_{j,i} \tag{10}$$

The torque $\mathbf{t}_{j,i}$ originating from various physical phenomena must be closed, which may benefit from the already known forces $\mathbf{f}_{j,i}$. The moment of inertia tensor $\mathbf{I}_i$ simplifies to an identity tensor for spherical particles, and hence can be pulled out of the time derivative in Eq. (10) for such simplified systems. Consequently, the above ordinary differential equation can be solved similar to that for translational motion.

However, an important note must be made with respect to Eq. (10): this equation is still valid, but less useful for nonspherical particles. This is since the tensor $\mathbf{I}_i$ characterizing the moment of inertia becomes time dependent in the global coordinate system. It is hence more convenient to track rotational particle motion in a frame of reference that is aligned with the (current) principal axes of inertia of each particle. In such a reference frame the moment of inertia tensor ($\mathbf{I}_i'$) is constant and diagonal. Therefore, one can exploit the following expression to relate time derivatives in a global and a body-fixed frame of reference:

$$d_t()|_{\text{global}} = d_{t'}()|_{\text{body}} + \boldsymbol{\omega}_i' \times () \tag{11}$$

Here the prime indicates quantities in the particle's frame of reference. Under the above assumption that the body frame of reference is aligned with the principal axes, one arrives at the celebrated Euler equations (Fan and Ahmadi, 1995):

$$\mathbf{I}_i' \cdot d_{t'}(\boldsymbol{\omega}_i') + \boldsymbol{\omega}_i' \times (\mathbf{I}_i' \cdot \boldsymbol{\omega}_i') = \sum_j \mathbf{t}_{j,i}' \tag{12}$$

This ordinary differential equation can be integrated in time in a straightforward manner. It remains to update the angular position of the particle, which is achieved most efficiently by time integration schemes that rely on quaternions (Lu et al., 2015).

2.3.5 Transport Equations for Macroscale Models

Transport equations for macroscale models are similar in form to their counterparts in mesoscale models (see Sections 2.3.2–2.3.4). Certainly, the individual terms in macroscale models describe very different physics, and in essence the following key differences have to be considered: (i) filtered quantities are considered in the balance equations and (ii) different closures need to be used for stresses and interphase exchange terms. The latter may require the solution of additional transport equations, e.g., to predict mesoscale velocity fluctuations in the gas and particle phase. A beautiful presentation of the resulting set of equations has been recently documented by Schneiderbauer (2017)—the interested reader is referred to this publication as a first read on the subject.

2.4 Governing Equations for Scalar Transport in the Continuous Phase

2.4.1 Particle-Resolved Approach

After accepting the assumption of a continuous phase that has constant physical properties (i.e., density, viscosity, heat capacity, etc.), the transport equation for a scalar quantity θ reads as:

$$\partial_t \theta^\circ + \nabla \cdot (\mathbf{u}^\circ \theta^\circ) = D_f \nabla^2 \theta^\circ + \mathbf{F}_\theta^\Omega. \tag{13}$$

In analogy to the presentation in Section 2.3.1, $\mathbf{F}_\theta^\Omega$ is a volumetric boundary forcing term that is used in some formulations (Municchi and Radl, 2017), and the superscript circle is used to identify point variables. D_f is a scalar diffusivity in the continuous phase, e.g., for thermal transport it is the thermal diffusivity $a_f = \lambda_f / (\rho_f c_{p,f})$, where λ and c_p are heat conductivity and heat capacity, respectively. Eq. (13) can be modified to consider homogeneous chemical reactions in a straightforward manner, given the flow, temperature, and concentration field are well resolved.

Frequently used boundary conditions are the classical fixed value or fixed flux condition at the particle surface, as well as specialized conditions for periodic boxes (Municchi and Radl, 2017; Sun et al., 2015).

Finally, it should be noted that in case appreciable changes of the fluid physical properties occur (e.g., the fluid density is affected by temperature, pressure, or species concentration), both momentum and scalar transport equations have to be extended (Dixon, 2017). While such an extension is relatively straightforward, it typically introduces a large number of additional dimensionless parameters that need to be considered.

In some applications (e.g., when modelling the droplet concentration in a gas; Askarishahi et al., 2017), it helps to view the scalar quantity to be tracked as a mass loading (i.e., consider the mass of scalar per unit mass of carrier fluid phase). Then, and when accepting a fixed density of the carrier phase, the scalar transport equation for the mass loading (i.e., a transport equation similar to Eq. 13) is exact in the sense that it accounts for density variations due to changes in the scalar concentration.

2.4.2 Particle-Unresolved Approach

Adopting the same assumptions as in Section 2.4.1, one arrives at the following transport equation for scalar local mean quantities:

$$\partial_t\left(\phi_f\theta\right) + \nabla\cdot\left(\phi_f\mathbf{u}\theta\right) = \nabla\cdot\left[\mathbf{D}_{\mathrm{eff},f}\cdot\nabla\left(\phi_f\theta\right)\right] + \mathbf{\Phi}_{\theta,p\to f}. \tag{14}$$

The most significant change compared to the particle-resolved approach is that an effective diffusivity tensor $\mathbf{D}_{\mathrm{eff}}$ must be considered instead of a scalar molecular diffusivity. Only in some simple situations this tensor may mutate to a scalar effective diffusivity. Corresponding closure models for the effective diffusivity—to be used in fixed bed applications—are well documented in literature (Delgado, 2006). In contrast, literature on the effect of particle relative motion on the effective diffusivity is rather scarce (Derksen, 2008a, b). As expected and shown by Derksen (2009), particle relative motion in fluidized beds accelerates scalar dispersion.

The volumetric scalar interface exchange rate $\mathbf{\Phi}_{\theta,p\to f}$ is modeled in analogy to the volumetric fluid–particle interaction force. Closure models for $\mathbf{\Phi}_{\theta,p\to f}$, useful to study heat and mass transfer, are summarized in Section 2.5.

Finally, a filtered version of the above scalar transport equation can be derived (Agrawal et al., 2013), a procedure which is formally identical to that for the momentum equation (see Section 2.3.5). The same applies to systems with heterogeneous chemical reactions (Holloway and Sundaresan, 2012). Similar challenges as for momentum transport arise when seeking for closures to the filtered scalar transport equation: the interphase exchange term needs to be modified, and an additional dispersion term has to be included. These challenges are currently about to be addressed in ongoing research activities (Cloete et al., 2017b, under review).

2.5 Closure Laws Required at the Mesolevel

2.5.1 Momentum Transfer: Drag and Buoyancy

2.5.1.1 Splitting the Fluid–Particle Interaction Force

Next, the most commonly employed splitting of the total fluid–particle interaction force $\mathbf{f}_{f \to p}$ is discussed. Note, this force is simply the negative of the particle–fluid force $\mathbf{f}_{p \to f}$, which was already introduced in Section 2.3.2.

In dense gas–particle systems a mean pressure gradient $\nabla \mathrm{p}$—generally pointing against the main flow direction—arises next to small-scale variations of the fluid-phase stress. The latter originate from the difference between the point variable p° (i.e., the local pressure, see Section 2.3.1) and its local mean counterpart p, as well as the local flow field $\mathbf{u}^{\circ}$. Hence, it is intuitively clear that it will be useful to account for two contributions to the total fluid–particle interaction force $\mathbf{f}_{f \to p}$: (i) one originating from a mean pressure gradient $\nabla \mathrm{p}$ and (ii) one accounting for the remaining variations in the fluid-phase stress.

The mean pressure gradient developing in dense gas–particle suspensions arises naturally when fluid is pushed through the interstices of the particles—this gradient can be readily extracted from PR-DNS. Most important, this pressure gradient is also known in Euler–Euler and PU-EL-based simulations. Hence, one will use this known information on the mean pressure gradient $\nabla \mathrm{p}$ to calculate its contribution to the fluid–particle force in EE or PU-EL simulations (this contribution is typically referred to as buoyancy or generalized buoyancy force). Consequently, a meaningful strategy is splitting off the contribution of the mean pressure gradient in PR-DNS from $\mathbf{f}_{f \to p}$ to isolate a remaining force contribution (see Section 4.2.3 for more details regarding this force calculation procedure). Here, the latter force contribution is called the drag force $\mathbf{f}_{\mathrm{drag},i}$ and needs to be reformulated in a closed form for the governing equations. In the context of PU-EL, i.e., when considering a particle with index i, the force splitting strategy can be formally expressed as:

$$\mathbf{f}_{f \to p,i} = -V_i \nabla p + \mathbf{f}_{\mathrm{drag},i} \tag{15}$$

We note in passing that in most dense suspension flows, the drag and mean pressure gradient force dominate fluid–particle interactions—contributions that may be characterized as a lift or added mass force are typically of lower importance. Thus, there is typically no need to perform a further splitting of the remaining force contribution, implying that $\mathbf{f}_{\mathrm{drag},i}$ lumps all fluid–particle interaction forces that originate from small-scale variations of the fluid-phase stresses into a single term.

2.5.1.2 Ensemble-Average vs Per-Particle

Certainly, closure models for the drag force were in the focus of previous developments, with the most prominent examples being the relations proposed by Wen and Yu (1966), Sundaresan and coworkers (Holloway et al., 2010, 2012; Yin and Sundaresan, 2009), or van der Hoef and coworkers (Beetstra et al., 2007; van der Hoef et al., 2005). The closure models developed by the latter two groups claim to be suitable for polydisperse suspensions and provide expressions for the *ensemble-average* drag force acting on each particle species (i.e., a class of particles that shares the same size).

A more recent study by Kriebitzsch et al. (2013) considered the fluctuation of the total per-particle fluid–particle force (i.e., the drag force plus the pressure gradient force). These authors found that the average force is only a poor representation of the distribution of forces. Clearly, per-particle and ensemble-average drag coefficient vary appreciably, leading to the need for a closer examination of how to split individual force contributions (Municchi and Radl, 2017). Also, the "drag like" contributions to the fluid–particle interaction force—explored just recently by Holloway et al. (2010)—focused only on average forces and no closure model for per-particle forces was presented. A similar gap in knowledge can be identified when considering lift forces (both for their average value and the per-particle distribution): while those forces are well understood in dilute suspensions—even for nonspherical particles (Amini et al., 2014; Di Carlo, 2009; Hölzer and Sommerfeld, 2009; Ouchene et al., 2015, 2016; Stan et al., 2013; Zastawny et al., 2012)—this is clearly not the case for dense suspensions (an exception are recent attempts by a French group; Ouchene et al., 2014).

Next, a classical description of drag forces acting on an ensemble of particles is presented. More recent developments are summarized in Sections 5.1–5.4.

2.5.1.3 Drag Force

In the context of suspension mechanics, the drag force acting on particles of species i is calculated under the assumptions that all neighboring particles move with the same speed. Thus, a "fixed bed" ansatz is followed, which reads:

$$\mathbf{f}_{\text{drag, fixed},i} = V_{p,i}\beta_{p,i}(\mathbf{u}_i - \mathbf{v}_i). \tag{16}$$

Here $\beta_{p,i}$ is a particle friction coefficient, which is different from the volume-specific friction coefficient β_i used in the context of EE models.

$V_{p,i}$ is a particle volume. A classical model for the particle friction coefficient in a monodisperse (fixed) bed of particles is (Beetstra et al., 2007):

$$\beta_{p,i} = \frac{18\mu_f}{d_p^2}\left(1-\phi_p\right)[F_0 + G_0], \tag{17}$$

$$F_0 = \frac{10\phi_p}{\left(1-\phi_p\right)^2} + \left(1-\phi_p\right)^2\left(1+1.5\sqrt{\phi_p}\right), \tag{18}$$

$$G_0 = \frac{0.413 Re_p}{24\left(1-\phi_p\right)^2}\left[\frac{\left(1-\phi_p\right)^{-1} + 3\phi_p\left(1-\phi_p\right) + 8.4 Re_p^{-0.343}}{1 + 10^{3\phi_p} Re_p^{-(1+4\phi_p)/2}}\right], \tag{19}$$

$$Re_p = \frac{\left(1-\phi_p\right)\rho_f d_p |\mathbf{u}_i - \mathbf{v}_i|}{\mu_f}. \tag{20}$$

Here F_0 and G_0 are dimensionless functions, and Re_p is a particle Reynolds number. For polydisperse (fixed) particle beds the situation becomes more complicated, and the following has been proposed (Beetstra et al., 2007):

$$\beta_{p,\text{poly},i} = \frac{18\mu_f}{d_p^2}\left(1-\phi_p\right)F_{\text{poly},i}, \tag{21}$$

$$F_{\text{poly},i} = [F_0 + G_0]\left[\left(1-\phi_p\right)y_i + \phi_p y_i^2 + 0.064\left(1-\phi_p\right)^3\right]. \tag{22}$$

Here y_i is a dimensionless particle size. Note, that in Eqs. (21) and (22) the particle Reynolds number used to compute F_0 and G_0 is defined slightly different than what is shown in Eq. (20). Sundaresan and coworkers (Holloway et al., 2010; Yin and Sundaresan, 2009) proposed another modification to compute drag forces in polydisperse suspensions. They follow the principal idea of Beetstra et al.'s model (Beetstra et al., 2007), but suggest:

$$\beta_{p,\text{YSH,poly},i} = \frac{18\mu_f}{d_p^2}\left(1-\phi_p\right)\left\{\frac{1}{1-\phi_p} + \left([F_0+G_0] - \frac{1}{1-\phi_p}\right)\left(a y_i + [1-a] y_i^2\right)\right\}, \tag{23}$$

$$a = 1 - 2.66\phi_p + 9.096\phi_p^2 - 11.338\phi_p^3. \tag{24}$$

Municchi and Radl (2017) find good agreement with Beetstra et al.'s closure model shown in Eq. (21). However, they propose a correction when applied to EL-based simulations (in order to be consistent with an EL model formulation):

$$F_{\mathrm{poly},MR,i} = -0.122 + 1.18 F_{\mathrm{poly},i} + 0.00352 F_{\mathrm{poly},i}^2. \tag{25}$$

Note, that all above models are valid only for the *ensemble-average* drag force, and strictly speaking not for individual particles (Municchi and Radl, 2017). Thus, when applied to EL-based models, only the average drag force—but not the particle-individual one—can be predicted correctly.

2.5.1.4 Generalized Buoyancy

Buoyancy forces arise due to gradients in the average stress field of the continuous phase. These forces act in addition to the forces originating from small-scale stress fluctuations, i.e., that captured by the drag force closures detailed in the previous paragraph. The generalized buoyancy force closure below accounts for the mean pressure and deviatoric stress field.

$$\mathbf{f}_{\mathrm{buo},i} = V_{p,i}\left(-\nabla p_i + \nabla \cdot \boldsymbol{\tau}_f\right). \tag{26}$$

2.5.2 Scalar Transport

Heat and mass transfer rates are extremely well explored in packed beds consisting of equally sized particles. This is true for both spherical and nonspherical particles (Singhal et al., 2017). Packed beds are examples for densely packed fluid–particle suspensions in which particles are not allowed to move. More scarce are closure models for less dense suspensions, which are often based on the correlation of Gunn (1978). Note, that these correlations are often only valid for Schmidt and Prandtl numbers close to unity. Also, consistency needs to be ensured: The work of Sun et al. (2015) highlights the importance of using the correct fluid temperature. For example, an inconsistency is caused when using the cup-mixing temperature when building a closure (e.g., from data collected in DNS), and a spatially averaged fluid temperature when applying the closure.

Closures for predicting heat and mass transfer rates in polydisperse suspensions are extremely scarce, same as information on the distribution of per-particle heat and mass transfer coefficients. Similarly, for high Reynolds, Schmidt, and Prandtl numbers, as well as high particle concentrations, literature is typically confined to a few studies (Derksen, 2014b). Clearly, a breakthrough in the sense of a rigorously valid correlation for the reliable prediction of heat and mass transfer coefficients is still missing.

2.5.2.1 Heat and Mass Transfer

The rate of heat and mass transfer is closed by correlations for the Nusselt (Nu) and Sherwood (Sh) number, respectively. Specifically, one writes the per-particle transferred amount of heat (and similarly for the mass transfer rate) as:

$$\dot{q}_T = Nu \lambda_f d_p \pi \left(T_{f,i} - T_{p,i} \right). \tag{27}$$

The most widely used expression for the Nusselt number in dense fluid–particle suspensions is (Gunn, 1978):

$$\begin{aligned} Nu = & \left(7 - 10\phi_f + 5\phi_f^2\right)\left(1 + 0.7 Re^{1/5} Pr^{1/3}\right) \\ & + \left(1.33 - 2.4\phi_f + 1.2\phi_f^2\right) Re^{7/10} Pr^{1/3} \end{aligned} \tag{28}$$

The appeal of Gunn's relation is that it satisfies four limiting relations (i.e., the low and high Reynolds number limit, the low Reynolds number single-particle limit, and the experimentally established dependencies of Nu with respect to the Prandtl and Reynolds number for the single-particle and packed-bed limit). Consequently, the shape of the relation shown in Eq. (28) has been widely accepted, and often only coefficients are refitted to new data (Deen et al., 2014; Singhal et al., 2017; Tavassoli et al., 2015).

Two important, but often neglected facts related to the application of Eq. (27) are that:

(i) the fluid's cup-mixing and local mean temperature are often confused (Municchi and Radl, 2017; Sun et al., 2015). Thus, care must be taken to ensure the correlation to be used is based on the local mean fluid temperature (or concentration) when applying EL models;

(ii) Available correlations are only able to predict ensemble-average Nusselt and Sherwood numbers. Consequently, one must accept typically large uncertainties when predicting per-particle transfer rates (Municchi and Radl, 2017).

Another detail—though relevant for mass transfer calculations involving wet granular materials only—is that of the available surface area. Often only a fraction of the particle's surface is covered with liquid, and only this wetted area must be considered for calculating mass transfer rates. A thorough description of how to calculate this wetted area fraction has been recently provided (Askarishahi et al., 2017).

All correlations presented above for Nu (and Sh) must be understood as particle- and ensemble-average dimensionless transfer rate coefficients.

Recently, an attempt to improve existing correlations was carried out (Kravets and Kruggel-Emden, 2017). These authors establish a correlation for the ensemble-average, but angle-resolved heat transfer coefficient in dense suspensions. "Angle-resolved" hereby refers to transfer coefficients that depend on a particle surface coordinate, specifically a parameterization using the angle between the mean flow velocity (i.e., $\mathbf{u}$) and a position on the particle's surface expressed in a spherical, particle-centered frame of reference. The useful aspect of the study remains questionable, since per-particle (but angle-averaged) heat transfer coefficients were not presented. However, the recent efforts documented in literature (Kravets and Kruggel-Emden, 2017; Municchi and Radl, 2017) stress the importance of future research to refine heat and mass transfer closure models to a state at which they are truly applicable to PU-EL-based models. Such efforts are discussed in greater detail in Sections 5.4.2 and 5.5.3.

2.5.2.2 Dispersion Coefficients

When considering heat and mass transfer, an additional closure naturally enters the system of equations to be used in a mesoscale model: the dispersion coefficient that characterizes how quickly an inert scalar is distributed randomly in the fluid domain. Delgado (2006) provides an excellent review of closures for packed beds. Dispersion rates of dissolved scalars in the fluid phase of an agitated suspension have been simulated only recently (Derksen, 2008b). A more recent study considered freely sedimenting suspensions, and additionally analyzed particle self-diffusivities in these systems (Derksen, 2014a). Both studies were performed for large Schmidt numbers and revealed the key scaling of dispersion coefficients in (sedimenting) suspensions, as well as their anisotropic nature. In summary, the development of closures for dispersion coefficients offers a similarly wide playground compared to that offered by heat and mass transfer rates. Clearly, extensions to (i) polydisperse systems, as well as (ii) systems made of nonspherical particles would be appreciated to provide more general closure models for dispersion coefficients.

2.6 Closure Laws Required at the Macrolevel

2.6.1 Phenomena That Affect Transport on the Macrolevel

The spontaneous clustering of freely sedimenting suspensions is characterized by strong fluctuations of the particle concentration and velocity. Fox and coworkers (Capecelatro et al., 2014, 2015) refer to this clustering and its consequences as "cluster-induced turbulence" (CIT). CIT is caused by the inherent instability of the fluidized state of a fluid–particle suspension

(Fullmer and Hrenya, 2017; Sundaresan, 2003). It appears that Jackson made the first attempt to probe the physical origin of CIT as early as 1963 (Jackson, 2000), and Glasser et al. (1998) were the first that simulated the instability that causes CIT with a computer. CIT does also occur in low Reynolds number suspensions (i.e., in systems where the particle-based Reynolds number is $O(1)$ or less). Most important, spontaneous clustering and CIT are fundamentally different from clustering of inertial particles that is induced by turbulent motion (Calzavarini et al., 2008). This is why one should not connect CIT with classical turbulence theory, but rather with spontaneous clustering phenomena.

CIT has been predicted with Euler–Euler (i.e., continuum-based) models, Euler–Lagrange models (Capecelatro et al., 2015; Radl and Sundaresan, 2014; Radl et al., 2012) (see Fig. 5 for an illustration), as well as PR-DNS (Derksen and Sundaresan, 2007; Uhlmann and Doychev, 2014). For dilute (i.e., 0.5 volume percent) liquid–particle systems, the density ratio and the Galileo number characterize the tendency for clustering (Uhlmann and Doychev, 2014).

2.6.2 Momentum Transfer

The primary effect of CIT is that it significantly alters momentum transfer rates between the fluid and the particles. In dilute suspensions only a weak

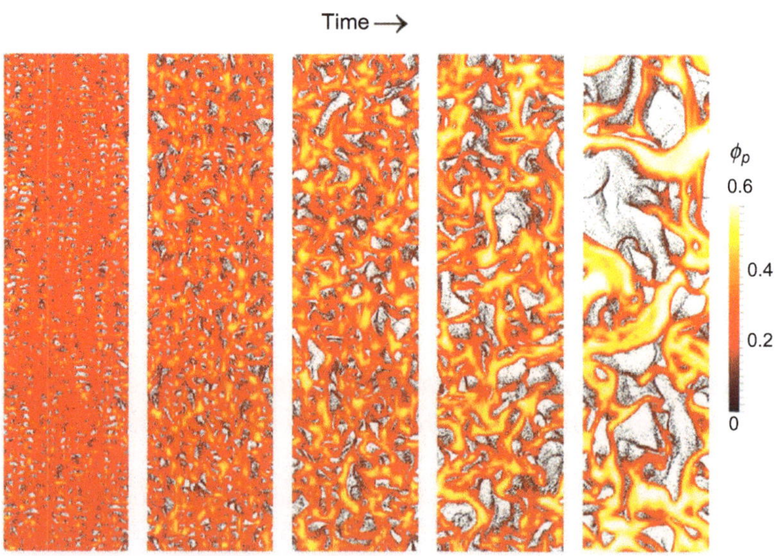

Fig. 5 Temporal evolution of the local particle volume fraction ϕ_p in a freely sedimenting gas–particle suspension (pseudo two-dimensional simulation using an Euler–Lagrange approach).

increase (i.e., approximately 12%) of the mean fluid–particle slip velocity is observed (Uhlmann and Doychev, 2014). Clustering causes a much more drastic increase (i.e., several hundred percent!) of the slip velocity for intermediately dense to dense suspensions. The latter can be characterized by a decrease of the average drag coefficient of up to 95% (Radl and Sundaresan, 2014). Clearly, there is a strong need to account for this decrease of the average momentum exchange coefficient, which is accommodated by correcting the drag coefficient. The principal shape of such drag modifications is:

$$\overline{\beta}_p = \beta_p \left[1 - H\left(\overline{\phi}_p, \Delta, \overline{\xi}\right)\right]. \tag{29}$$

Here H is a (typically scalar-valued) correction function, and $\overline{\xi}$ is a placeholder for additional markers that are available on the level of filtered quantities and affect the correction H. A large collection of potential markers has been tested in literature—mainly by applying filtering tools to cluster-resolved simulation data—with no ultimate consensus so far (Cloete et al., 2017a,b; Igci and Sundaresan, 2011; Milioli et al., 2013; Ozel et al., 2013; Schneiderbauer, 2017).

Correction functions H for the drag have been carefully analyzed for monodisperse gas–particle suspensions over the past 10 years. Concerning polydisperse systems, some pioneering work has been done by the group of Sundaresan on bidisperse systems (Holloway and Sundaresan, 2014). This group attempted to shed light on the effect of clustering on effective momentum exchange coefficients in sedimenting suspensions, as well as the effective stress. This work also established the framework for a filtered multifluid model (in the context of EE models), that might help to explain segregation phenomena in large-scale devices. Follow-up work following this thought has been already published (Zhou and Wang, 2015). In summary, what has been done for polydisperse suspensions is still far from being sufficient.

2.6.3 Stress in the Particulate Phase

Closures that account for the effect of mesoscale phenomena on the particle- and gas-phase stresses are required on the macrolevel (Agrawal et al., 2001; Igci et al., 2008; Sarkar et al., 2016; Schneiderbauer, 2017). In addition to these "mesostresses," the filtered kinetic theory-based stresses, as well as the filtered frictional stresses, potentially, must be considered.

Several assumptions are generally adopted to limit the number of closures in modelling mesoscale stresses:

(i) fluid-phase mesoscale stresses are assumed to be small. This is valid as long as the fluid's density is much smaller than that of the particles

(ii) contributions from the filtered kinetic theory-based stresses—relative to the mesoscale solids stresses—decrease with increasing filter size. Thus, for models on the macrolevel that aim on industrial scale applications, filtered kinetic theory-based stresses can be neglected

(iii) filtered frictional stresses are relevant only for densely packed regions and have been given a lower amount of attention with only a few exceptions (Schneiderbauer and Pirker, 2014).

Mesoscale stresses developing in the particulate phase are expected to be the second most important piece of missing information that needs closure in models on the macrolevel. Indeed, a recent study Schneiderbauer (2017) reported that forces due to mesoscale (particle) stresses may even exceed that caused by drag. This clearly indicates that forces due to mesoscale stresses, although smaller than the filtered drag forces on average, are significant. A typical Boussinesq-like ansatz for closing the particle-phase mesoscale stresses in a macroscale model is (Schneiderbauer, 2017):

$$\boldsymbol{\tau}_{p,\mathrm{meso}} = \mathbf{I}\frac{2}{3}\overline{\phi}_p\rho_p k_p - 2\overline{\phi}_p\mu_{p,\mathrm{meso}}\overline{\mathbf{S}}_p. \tag{30}$$

The reader should be reminded that an overbar indicates filtered quantities (which are explicitly calculated in the macromodel). Also, $\mathbf{I}$ is the identity matrix, and $\mathbf{S}$ is the deviatoric part of the rate of deformation tensor. k_p (i.e., the kinetic energy of mesoscale velocity fluctuations) and $\mu_{p,\mathrm{meso}}$ (i.e., the mesoscale viscosity) require closure, for which a variety of options exist. An especially appealing option is considering a balance equation for k_p. Thus, one first aims at predicting mesoscale velocity fluctuations to close the pressure-like term in the expression for the mesoscale stress. Subsequently, the viscosity can be closed using a simple mixing length assumption. Also, knowing k_p (and k_g, i.e., the gas phase mesoscale kinetic energy) greatly helps to close the coefficient for interphase momentum transfer in macroscale models (Schneiderbauer, 2017).

This approach to close the stress—while certainly at the bleeding edge of development—has the shortcoming that it is inherently assumed that the mesoscale stress state can be characterized with two scalar quantities: a pressure-like term and a viscosity term. This implies isotropy in the subgrid scale velocity fluctuations (and hence implies the absence of normal stress differences), which may not be justified. In Sections 6.2 and 6.4 some recent efforts are documented that make an attempt to overcome this shortcoming.

2.6.4 Heat and Mass Transfer

Similar to what is done for the drag, exchange coefficients for heat and mass, as well as heterogeneous reaction rates need to be corrected in case CIT occurs (Agrawal et al., 2013; Cloete et al., 2011; Holloway and Sundaresan, 2011, 2012; Schneiderbauer, 2017; Sundaresan et al., 2013). Modifications to the heat and mass exchange coefficients (denoted here as γ) are typically closed following the ansatz (Agrawal et al., 2013):

$$\overline{\gamma} = \gamma\left(1 - Q\left(\overline{\phi}_p, \Delta\right)\right). \tag{31}$$

Here Q is a correction function that is typically taken to depend on the filtered particle volume fraction and the filter size. However, expansion to other sets of markers can be done similar to what has been discussed in Section 2.6.2.

In analogy to mesoscale stresses in the momentum equations, a dispersive flux due to unresolved velocity and temperature (or species concentration) fluctuations must be considered. This dispersion flux is modeled similar to a diffusion term, with a dispersion coefficient D approximated by a Smagorinsky-like ansatz (Agrawal et al., 2013; Cloete et al., 2017b). For example, for the particulate phase one can model this mesoscale dispersion with:

$$D_{p,\text{meso}} = C_{p,\text{meso}}\left(\overline{\phi}_p\right)\Delta^2 \left\|\overline{\mathbf{S}}_p\right\|. \tag{32}$$

Here $C_{p,meso}$ is a Smagorinsky-like parameter that is modeled as a function of the filtered particle volume fraction (Agrawal et al., 2013).

3. ALGORITHMIC ASPECTS

3.1 Flow Simulators

Simulations of granular and suspension flows started in the late 1970s, a time where physical chemists had already invented the widely used molecular dynamics simulation approach. Specifically, the soft-sphere simulation method (Cundall and Strack, 1979) was published in 1979. Some earlier publications of Cundall in this matter even date back to 1971. This method is now widely referred to as the "discrete element method", or simply DEM. Nearly at the same time, the second ingredient for modern computational granular rheology, i.e., the Lees–Edwards boundary conditions were published (Lees and Edwards, 1972). Approximately 10 years later, Stokesian dynamics (SD) for studying suspension flows at low Reynolds numbers was developed (Brady and Bossis, 1988; Sierou and Brady, 2001). SD was a key tool for generating new insight with limited computational resources since it does not require a discretization of the fluid domain. This study nicely illustrates the desperate need for more computational resources that was clearly limiting scientific progress at that time.

What followed was a fast development of tools in the 1990s: one example is the Lattice Boltzmann method (LBM), which was first applied by Ladd (1994) to model suspension flows. Starting around the year 2000, LBM was increasingly used in the community as the data generation workhorse, as an example see the work of Rohde et al. (2002) or Derksen (2008b). In addition, conventional finite volume techniques (e.g., IBM, see Fig. 6 for an example) are now extensively used and progressively refined (Municchi and Radl, 2017; Podlozhnyuk et al., 2016; Sun et al., 2015; Uhlmann and Doychev, 2014).

The above simulators are used to solve microscale models—all details of the flow system on the scale of individual particles are resolved. A plethora of commercial and open-source simulators exists that implement various computational strategies (see Section 1.3) to evaluate "engineering" models (i.e., EE-based or PU-EL-based models). A more comprehensive summary of such strategies is provided in various review articles (e.g., Section "Mechanistic Process Modelling" of Rantanen and Khinast (2015), as well as Fullmer and Hrenya (2017).

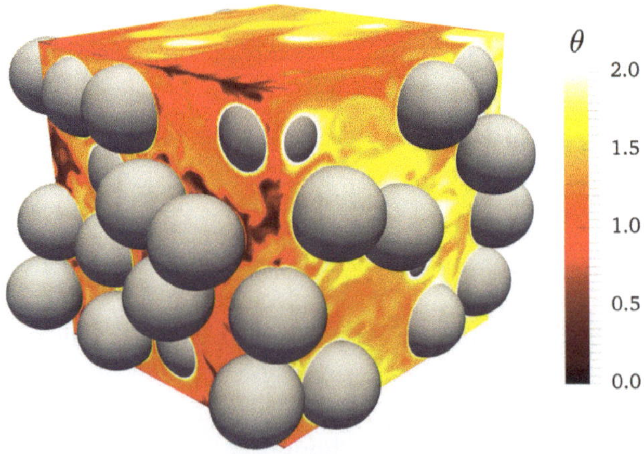

Fig. 6 Temperature field θ in the fluid phase of a gas–particle suspension bound by two infinite parallel adiabatic walls (the Reynolds number is 400, the Prandtl number is 1, and the particle volume fraction is 0.4).

3.2 Filtering Tools

3.2.1 Requirements and Challenges Associated With Spatial Filtering

At the most fundamental level, PR-DNSs are used to derive coefficients for heat, momentum, and mass transfer in dense particulate systems (Deen et al., 2012, 2014; Feng and Musong, 2014; Tavassoli et al., 2013; Zaidi et al., 2014). Typically, a certain number of realizations for the case studied is needed in order to derive statistically meaningful correlations (Tenneti and Subramaniam, 2014 ; Tenneti et al., 2013). This approach requires to process data from large data sets in order to compute averaged (mean) quantities (which are needed to evaluate transfer coefficient), standard deviations, or other statistics like the distribution of the angle between two vector fields (Kriebitzsch et al., 2013). Also, strategies that rely on the temporal evolution of flow quantities (e.g., the transients of one-dimensional concentration profiles) are in use to evaluate transport coefficients in fluid–particle systems. An example of such a strategy is the work of Derksen, who quantified scalar dispersion rates in sedimenting liquid–particle suspensions (Derksen, 2014b).

The same approach, i.e., considering statistical data of, e.g., the velocity fluctuations, can be used on intermediate length scales when deriving meso- and macroscale models for engineering applications. Typically, this results in an Eulerian grid coarsening approach, e.g., by deriving models for the subgrid-scale fluid–particle agitation (Igci et al., 2008). Favre averaging of relevant fluid variables (e.g., the fluid velocity), and fluid–particle interactions (e.g., the coupling force) is generally adopted to derive these closures. A Favre average refers to the phase mass-based average as detailed in Appendix A.1, and is in its standard form defined as $\widetilde{(*)} = \overline{\phi(*)}/\overline{\phi}$ where ϕ is the phase fraction.

In case an Euler–Lagrange approach is followed, the effect of particle coarsening (i.e., each simulated particle is a proxy for a prescribed number of particles named parcel) has to be taken into account as well (Ozel et al., 2016). All these examples document current challenges that need to be met by a universal filtering tool.

In principle, the application of an appropriate filtering strategy is straightforward once the fluid–particle flow simulator is available. However, filtering of scientific data and coarse-graining poses several challenges from the software point of view. For example, spatio-temporal averages have to be computed across different processors for the (typically large) filter sizes. Typically, filter sizes to be used when filtering PR-DNS data have a size of two to five particle diameters. Thus, filtering is typically performed over 20^3 to 50^3 Eulerian grid cells, often located on different processors. This requires an

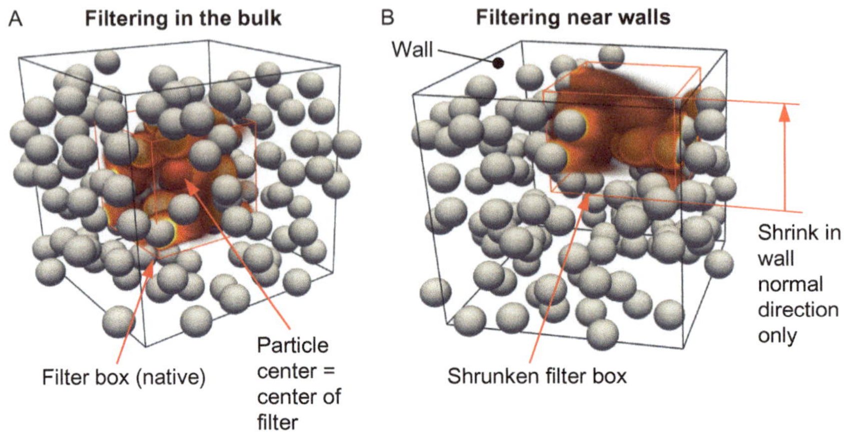

Fig. 7 Illustration of the shape and size of the filtering region in the bulk (A), as well as in the proximity of walls (B).

algorithm that can deal with parallel communication, and does not require mirroring the full field information on every processor. The latter is of course a feasible approach, however, when aiming on large-scale simulations this would require an excessive use of RAM. At the same time, parallel communication of local field values requires significant network resources due to the large amount of data to communicate. Another issue is the amount of data generated during the simulation run: the implemented algorithms should be able to work on the fly in order to process (i.e., time-average) data from different time steps. Also, there should be a clear separation (in terms of namespaces and classes) between the simulator and the postprocessing utility such that the latter can be linked to different simulators.

Finally, the library should be modular in order to make the addition of new features as easy as possible. For example, filtering near walls might require changing the filter box shape and size (see Fig. 7 for an illustration).

3.2.2 The Tool CPPPO

The library CPPPO addresses the above challenges. While most of the existing filtering algorithms documented in literature were developed for image processing applications (Nakariyakul, 2013; Tsai and Yeh, 2012), CPPPO is able to handle three-dimensional data sets in parallel. Specifically, CPPPO can process data sets from complex geometries and unstructured meshes. CPPPO features a flexible code architecture tailored for scientific computations on high-performance clusters. The library is designed to perform three kinds of operations on the data set:

1) Filtering: field volume averaging that can be performed on every cell (for the Eulerian filter option) or at specific user-defined locations (for the Lagrangian filter option). The user can customize the kernel function by adding an arbitrary amount of weights (which have to be scalar fields), or by modifying its form in general.
2) Sampling: This operation allows to take samples from the domain (results from the filtering operations may be sampled as well) and relate each sampled value with one or more markers. For example, CPPPO can sample a spatially filtered fluid velocity field at every cell using the fluid-phase fraction as marker. Typical sampling operations include: (i) General sampling: This routine draws samples over the whole domain, or just a portion of it. Sampled fields and markers are defined by the user. Also, general sampling allows the use of a formula parser implemented in CPPPO to draw samples of quantities which are not explicitly calculated in the simulator. These formulas include instructions to compute the gradient, the divergence, dot and double-dot products, as well as simple algebraic manipulations. (ii) Angle vector–vector: This routine can sample vector fields using the angle between the original and a second vector field as marker. (iii) Two point correlation: This routine will sample the value of the trace of a two point correlation.
3) Binning: Data collected from sampling operations can be collapsed using binning operations. The marker field values are discretized according to the user input, and a conditional averaging calculation is performed on the sampled field. This data collapsing allows reducing the amount of data that needs to be written to disk. Binning is useful in case the user is only interested in correlations between the means of the sampled quantities and one (or more) markers.

For every user-specified filter (i.e., kernel function), the library performs these three operations in sequence (see Fig. 8). This loop is referred to as the "FSB loop" (Filtering, Sampling, and Binning).

CPPPO can be run in parallel with any simulator that splits the computational domain in several box-shaped subdomains. Since CPPPO is designed for spatial filtering, data should be decomposed according to their position in space. This is true in almost the totality of currently available simulators for CFD either using a finite volume approach (e.g., OpenFOAM®, ANSYS FLUENT®, *Code_Saturne*®, STAR-CCM+®, AVL FIRE®, etc.) or not (e.g., Palabos®, Nektar++®, Nek5000®, etc.). Parallel communication in CPPPO mainly relies on collective MPI operations, since most of the time all processors have to synchronize during the calculations.

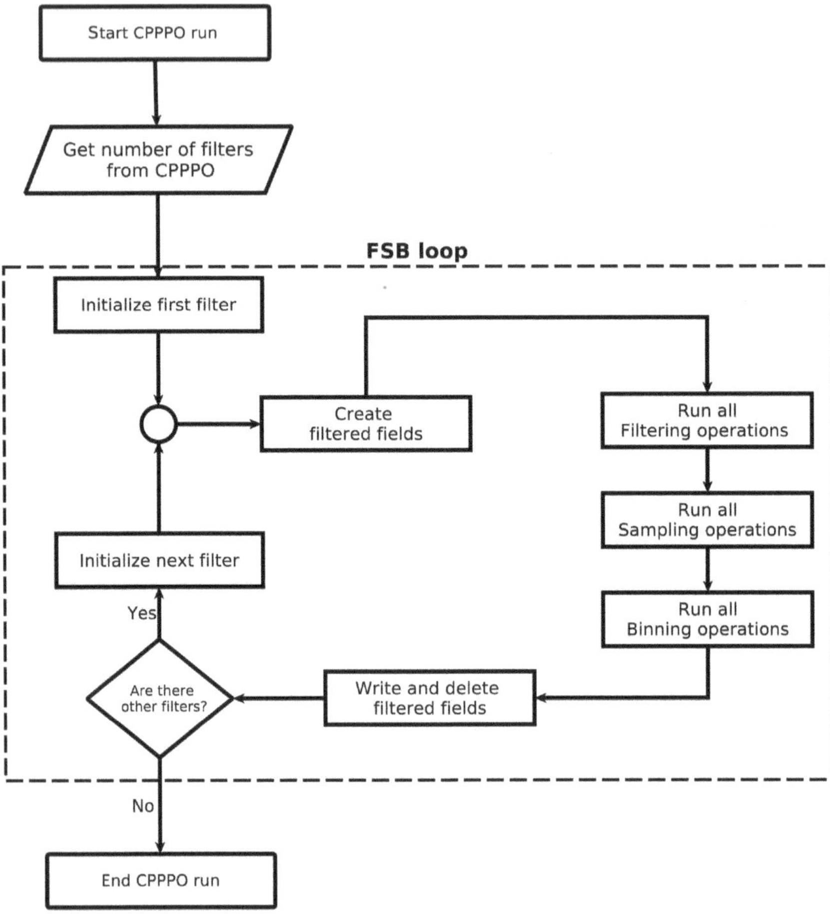

Fig. 8 The Filtering–Sampling–Binning loop implemented in the filtering tool "CPPPO." *Reproduced from Municchi F, Goniva C, Radl S: Highly efficient spatial data filtering in parallel using the opensource library CPPPO,* Comput Phys Commun *207:400–414, 2016.*

These MPI routines have shown excellent performance in many applications (Ma et al., 2013; Mamidala et al., 2008) on HPC hardware. The load partitioning in parallel computations with CPPPO is mainly a function of the domain decomposition and the distribution of sampling locations. Hence, it is mostly user dependent. This is particularly true when using Lagrangian filtering operations (i.e., filtering is performed at predefined sampling positions). The reason is that the user affects directly the processor load in such a situation. For example, in case all sampling positions are positioned in a subdomain belonging to the same processor, the calculation would be slow. Thus, all operations would be focused on just one processor.

More details related to the coupling of CPPPO to flow simulators, as well as the implementation of cell selection and parallel filtering routines are provided in a recent publication (Municchi et al., 2016).

3.2.3 Strategies for Computing Averages in Parallel

Since the calculation of filtered quantities implies long range data exchange operations, processor communication has to be taken into account when designing an algorithm to numerically evaluate the corresponding integrals. On a discrete level these integrals transform to a summation, which can be written as:

$$\overline{\phi}_i = \sum_{j=0}^{j=N_f} V_i \phi_i / V_f. \tag{33}$$

Where the sum is over all N_f cells inside the filter region, V_i is the volume of the i-th cell, and V_f is the total filter volume. The extension of the above equation to Favre averaging or arbitrary weighted averaging is obvious. The above calculation has to be performed for every cell i in order to compute a complete field of the filtered quantity. Also, it is necessary to evaluate which cells are inside the filter (i.e., a cell selection is necessary) before the calculation can be started.

The approach described by Eq. (33) is called the *convergent approach* for filtering. This is because, after the list of cells inside the filter is assembled, data from the neighboring cells are passed to the location where the filter is centered. This approach requires (for every location to filter) $N_f - 1$ summation and multiplication operations, and one division operation. The amount of multiplication operations could be reduced in case the multiplied field values are stored and then communicated. However, this would require additional memory. Also, communication of the cell list and values with other processors has to be performed. As shown in Fig. 9 (left panel), the convergent approach requires the communication of every required cell data owned by another processor. In our case, three values need to be transferred from processor 2 to processor 1. The convergent approach is the most basic approach for spatial filtering, and most of the available filtering algorithms used for image processing are based on it.

In order to reduce the computational load and enhance parallel efficiency, a novel approach named the *divergent approach* was developed. The divergent algorithm does not evaluate the filtered value at any position

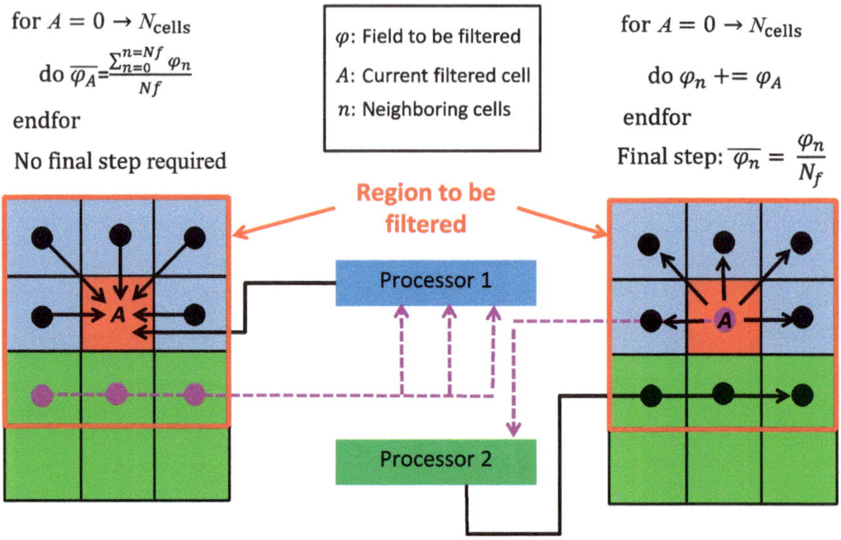

Fig. 9 Convergent approach (*left*) and divergent approach (*right*) for spatial filtering. *Continuous arrows* represent intraprocessor operations, while *dashed arrows* indicate data exchange between processors. Processor domains are identified with the owner color, *red cells* represent the current cell to be filtered. *Dots* represent cell centres involved in local data operations (*black*) and parallel data operations (*purple*). N_{cells} is the total number of cells per processor, and N_f is the number of cells within the region to be filtered.

sequentially, but updates the filtered fields at every step, and ends with a final division step. Specifically, every step consists of:

i. selecting a cell from the computational domain.

ii. creating a list of cells located in the region to be filtered around the selected cell.

iii. multiplying the field value at the selected cell with the required weight (i.e., cell volume or mass density). *Note*: when the filter size tends to the domain size only one cell value needs to be stored and communicated instead of having to allocate and communicate the whole field.

iv. communicating and adding the multiplied field values to all cells inside the cell list.

The loop has to be repeated for every cell inside the domain. At the end, one last step is needed to divide the values of filtered fields by the filter volume (or by the summed weights in case of Favre averaging). This approach requires (for every cell) $N_f - 1$ additions, but only one multiplication. The number of divisions in the final step equals the total number of cells. Overall, less multiplication operations are required in the divergent approach compared to a convergent approach (without allocating memory for the multiplied fields as explained earlier). Most important, the key advantage of the divergent algorithm over the convergent algorithm is the amount of data that needs to be communicated. As shown in Fig. 9 (right panel), in the divergent algorithm the direction of communication is reversed, and just the field value at the current cell needs to be communicated once. The communicated value is then processed locally on the relevant processor (in our case processor 2), which does not involve any communication overhead.

It should be clear that the computational bottleneck for these kind of algorithms is not the number of standard operations, but the number of MPI operations. While image filtering algorithms tend toward a reduction on the number of standard operations, the algorithms implemented in CPPPO have the reduction of the number of MPI operations as the main goal. Since field and mesh data can be very large in terms of the consumed memory, it is often not feasible to rely on massive data copying and thus, processor communications are rather frequent.

4. VERIFICATION STRATEGIES

In what follows test cases for the verification of computational tools are summarized. Validation against experimental data is a topic of its own and has been followed quite successfully for a large array of models and closures. The work of Buist et al. (2017) is a representative example of a validation study; more details related to experimental validation strategies are suppressed for brevity in what follows.

4.1 Verification Strategy for Filtering Tools

The accuracy of a filtering tool can be tested by considering well-known problems of fluid dynamics: Stokes flow and irrotational (i.e., potential) flow around a sphere. The main objective of these tests is to evaluate the accuracy of the parallel filter routines and to illustrate the dependency of the results on the grid size. Therefore, CPPPO results where compared against analytical solutions of filtered quantities at the particle center (Municchi et al., 2016). Recalling the analytical solution for Stokes flow (i.e., zero Reynolds number) around a sphere with radius R, and when considering the velocity component u_x in the stream-wise (i.e., x-) direction, the flow field is described by:

$$u_{x,\text{Stokes}} = U_\infty \left[\cos^2\theta \left(1 + \frac{R^3}{r^3} - \frac{3R}{2r} \right) + \sin^2\theta \left(1 - \frac{R^3}{4r^3} - \frac{3R}{4r} \right) \right]. \quad (34)$$

Here, θ and r describe the polar and radial position in a spherical coordinate system, respectively (the solution is symmetric with respect to the azimuthal coordinate). U_∞ is the flow velocity far from the particle. The corresponding solution for irrotational flow (i.e., a flow characterized by an infinitely large Reynolds number) past a sphere is:

$$u_{x,\mathrm{Irr}} = U_\infty \left[\cos^2\theta \left(1 - \frac{R^3}{r^3}\right) + \sin^2\theta \left(1 + \frac{R^3}{2r^3}\right) \right]. \tag{35}$$

Consider now a spherical filter and define a dimensionless filter size as $\rho = R_f/R$, where R_f is the filter radius. Integration of the above equations to obtain the mean (i.e., denoted as $\overline{u}$), and the variance (corresponding to the subgrid stress tensor τ^{sgs}) of the stream-wise velocity component leads to:

$$\overline{u}_{x,\mathrm{Stokes}} = \frac{2\rho^2 - \rho - 1}{2(\rho^2 + \rho + 1)} U_\infty, \tag{36}$$

$$\tau^{\mathrm{sgs}}_{xx,\mathrm{Stokes}} = \frac{18\rho^5 - 32\rho^4 + 14\rho^3 - 3\rho^2 + 2\rho + 1}{5\rho^3(4\rho^4 + 8\rho^3 + 12\rho^{12} + 8\rho + 4)} U_\infty^2, \tag{37}$$

$$\overline{u}_{x,\mathrm{Irr}} = U_\infty, \tag{38}$$

$$\tau^{\mathrm{sgs}}_{xx,\mathrm{Irr}} = \frac{U_\infty^2}{5\rho^3}. \tag{39}$$

These relations provide a set of cases to verify filtering routines for the situation of a top-Hat filter kernel in spherical coordinates. Corresponding results produced with the tool CPPPO are summarized in Fig. 10 (Municchi et al., 2016).

4.2 Selected Verification Cases for Benchmarking Particle-Resolved DNS

Simple verification cases for PR-DNS include drag force and heat transfer from a single sphere. Analytical solutions exist for these cases. However, more relevant for suspension flow simulators are multiparticle configurations that probe all relevant details of the algorithm (e.g., handling of Eulerian cells that are intersected by multiple particle surface). Such cases are summarized next.

4.2.1 Creeping Flow Past a Periodic Regular Array of Spheres

To simulate flow past a periodic array of spheres, it is sufficient to simulate the flow past a single sphere in a fully periodic box. This test case allows evaluating the accuracy of the method when two or more particles are close or in contact.

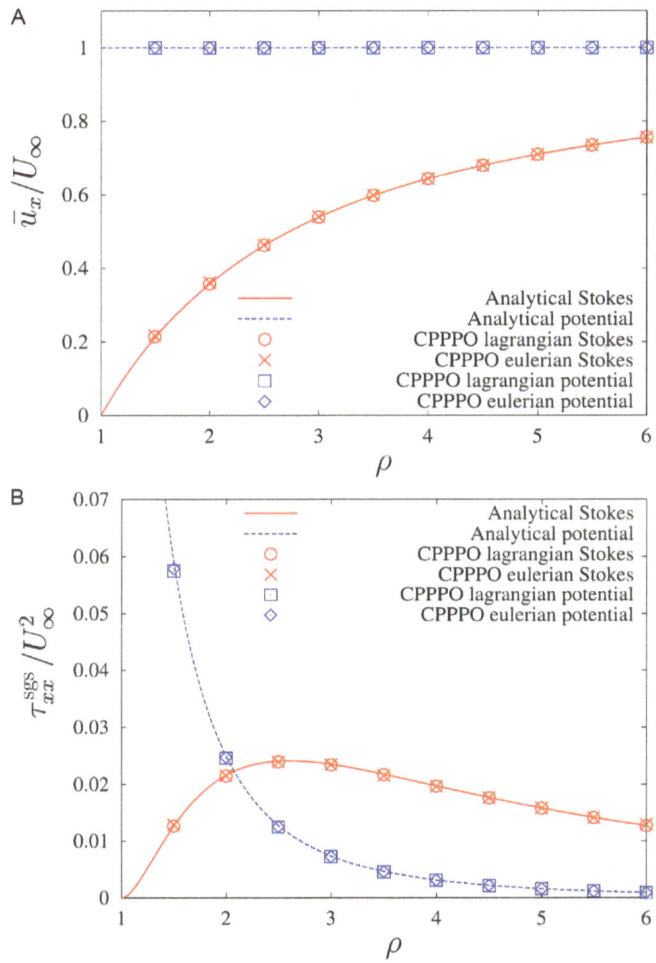

Fig. 10 Comparison between CPPPO's Lagrangian and Eulerian filtering tools with analytical results for the filtered quantities at the particle centre. Results for the Favre average of the velocity are shown in (A), while in (B) the Favre variance of the velocity was calculated.

Thus, one can assess the effect of a boundary layer reconstruction algorithm that adjusts its order according to the local situation (Municchi and Radl, 2017). For this case, the equations summarized in Section 2.3.1 were solved using a body force to drive a flow field such that the particle Reynolds number is 2×10^{-5} (Municchi and Radl, 2017). The particle diameter was varied for a fixed domain size to adjust the particle volume fraction. Predictions for the dimensionless force (i.e., the force measured from the simulation divided

Table 2 Relative Deviation of the Magnitude of the Dimensionless Drag Force F to the Analytical Results of Zick and Homsy (1982)

ϕ_p	$d_p/\Delta x$	$F_{\text{Zick–Homsy}}$	$\Delta F_{\text{Deen et al.}}$ (%)	$\Delta F_{\text{Municchi–Radl}}$ [%]
0.5236	32	42.14	1.3	1.1
0.450	30.4	28.1	−2.8	1.4
0.343	27.8	15.4	−1.2	2.7
0.216	23.8	7.44	−0.1	−0.8
0.125	19.9	4.292	−0.7	0.3
0.064	15.9	2.81	−1.3	−1.6
0.027	11.9	2.008	−1.9	−2.8

Δx is the mesh size. The settings are the same as that employed by Deen et al. (2012).

by the corresponding value for Stokes drag) can then be benchmarked with data from literature (see Table 2).

4.2.2 Three Particles Aligned With the Main Flow

The case of forced convection past a chain of three spheres is a classical verification case that allows benchmarking predictions for the per-particle Nusselt number and drag coefficient. Recent reference data for this case can be found in literature (Maheshwari et al., 2006; Tavassoli et al., 2013) for different values of the spacing between the three particles. The flow field predicted by our newly implemented PR-DNS solver (Municchi and Radl, 2017) is show in Fig. 11. This data, as well as our results for Nusselt number and drag closely match the above literature references.

4.2.3 Densely Packed Random Arrays of Particles

Ensemble-average fluid–particle transfer coefficients (e.g., for momentum and heat) have been widely studied by means of computer simulations for random arrays of particles due to their high industrial significance. Note, that these arrays were typically considered to be periodic and their microstructure to be perfectly isotropic. The setup procedure of such systems is detailed by van der Hoef et al. (2005). The large body of results being now available has been casted in an impressive amount of correlations documenting (i) drag coefficients, as well as (ii) Nusselt (or Sherwood) numbers (Deen et al., 2014; Municchi and Radl, 2017, 2018).

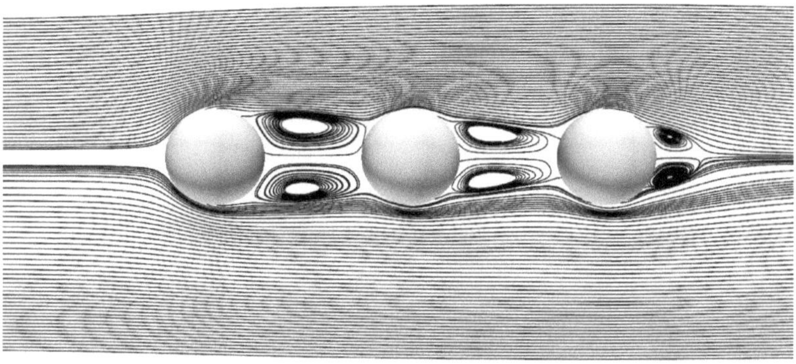

Fig. 11 Streamlines illustrating the flow field around a chain of particles that is aligned with the main flow direction.

4.2.3.1 A Note on the Drag Force Calculation

With respect to drag, and as outlined in Section 2.5.1, it is important to define the drag force in a meaningful way such that it is correctly separated from the total fluid–particle interaction force. Hence, the question arises how to account for the pressure contribution when computing drag forces from PR-DNS. The standard approach—widely adopted in literature (Beetstra et al., 2007; van der Hoef et al., 2005)—is to rescale the particle-average fluid–particle interaction force with the mean voidage. Alternatively, one can take a more fundamental route and subtract the per-particle buoyancy force (calculated from the mean pressure gradient experienced by an individual particle) from the fluid–particle interaction force. While both approaches yield identical drag forces in the case of a monodisperse bed of particles, differences do exist in case polydisperse beds are considered. The reasons behind this discrepancy have been highlighted just recently (Municchi and Radl, 2017). They are most obvious in case one considers a single large particle in an infinitely large bed of smaller particles: clearly, in this case the pressure gradient will be set by the small particles. Consequently, the buoyancy force experienced by the larger particle is solely defined by the fluid–particle interaction force of the smaller particles—the fluid–particle interaction force acting on the large particle is of no relevance to the pressure gradient. Thus, subtracting the pressure gradient force for the large particles must account in some way for the size of the smaller particles, which is not the case when using the standard approach. Hence, it is of outmost importance to clarify the exact definition of the drag fore during verification studies.

4.2.3.2 Results for Drag

Adopting the widely used standard approach, almost the totality of previous work found excellent agreement of the average drag force with the correlation provided in Section 2.5.1 (Beetstra et al., 2007). This is true for both mono- and bidisperse particle suspensions, and our more recent result (Municchi and Radl, 2017) also agree with this data. Most important, our results indicate that this agreement even holds in case the drag coefficient is evaluated for different filter sizes ρ (see Fig. 12).

4.2.3.3 Results for Heat Transfer

With respect to heat transfer correlations, a significant amount of reference data has accumulated in literature. In order to make a fair comparison, one should take care when comparing reported Nusselt numbers, since their exact definition matters. The key issue is the definition of the average fluid temperature, for which a (i) spatially filtered or a (ii) flux-averaged value can be used. The difference between these two choices has been carefully evaluated just recently (Sun et al., 2015). When accounting for this difference, our recent work shows that a refitted-version of the popular Gunn correlation (Deen et al., 2014) approximates our data for particle-average heat transfer coefficients well (see Fig. 13; for the largest filter size ρ the number of samples

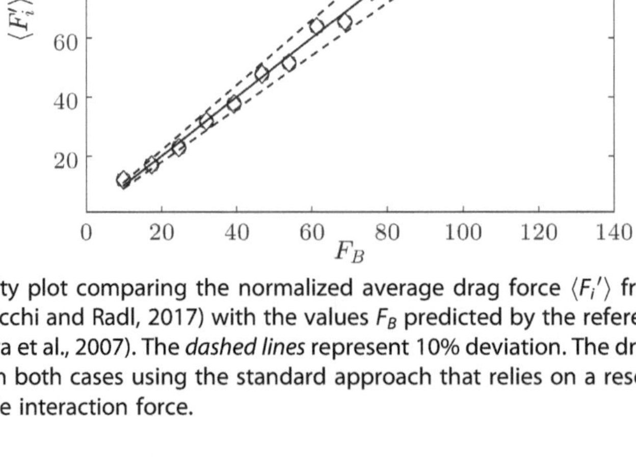

Fig. 12 Parity plot comparing the normalized average drag force $\langle F_i' \rangle$ from a recent study (Municchi and Radl, 2017) with the values F_B predicted by the reference correlation (Beetstra et al., 2007). The *dashed lines* represent 10% deviation. The drag force was calculated in both cases using the standard approach that relies on a rescaling of the fluid–particle interaction force.

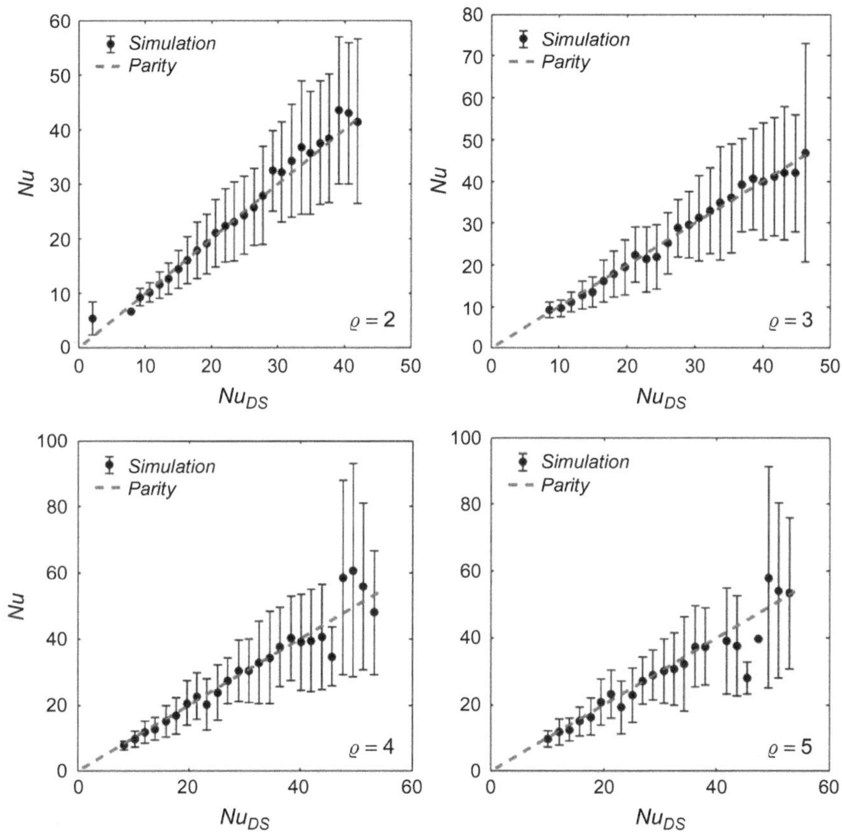

Fig. 13 Parity plot of average Nusselt number computed in our recent study (Municchi and Radl, 2018, averages are represented as *dots* and the *errors bars* extend to two standard deviations) and a corrected version of a popular correlation (Deen et al., 2014). The definition of the Nusselt number is based on filtered fluid temperatures. Different panels refer to different values of the dimensionless filter size ϱ.

to calculate statistics was lower due to computational reasons, Municchi and Radl, 2018). Thus, it appears that the currently most plausible correlation for the Nusselt number is:

$$Nu = N_{\mathrm{corr,Sun}} \left[\begin{array}{c} \left(7 - 10\phi_f + 5\phi_f^2\right)\left(1 + 0.17 Re^{1/5} Pr^{1/3}\right) \\ + \left(1.33 - 2.31\phi_f + 1.16\phi_f^2\right) Re^{7/10} Pr^{1/3} \end{array} \right], \tag{40}$$

$$N_{\mathrm{corr,Sun}} = \left[1 - 1.6\phi_p \phi_f - 3\phi_p \phi_f^4 \exp\left(-Re_p^{2/5} \phi_p\right)\right]^{-1}. \tag{41}$$

The above correlations can be used in case spatially filtered data for the fluid temperature are available, i.e., in case PU-EL-based models are used.

We note in passing that an alternative route for correlating heat transfer coefficients is based on relating them to drag coefficients, i.e., assuming an analogy between heat and momentum transfer. Indeed, our previous work (Municchi and Radl, 2017) indicated that such an analogy exists. Such an analogy approach is especially appealing in case per-particle drag coefficients are available, since it suggests that then also per-particle heat transfer coefficients can be calculated with ease.

5. RECENT DEVELOPMENTS FOR CLOSURES DESCRIBING TRANSFER RATES IN MESOSCALE MODELS

5.1 Effects Due to Differences in the Mean Particle Speed: The Fluid-Mediated Particle–Particle Drag

It remains to discuss the effect of relative particle motion on the drag force. This is since the closures detailed in Section 2.5.1 are only valid for situations in which all particles move with the same speed. In case particles move relative to each other, and the suspension is characterized by a high Stokes number and low Reynolds number, Yin and Sundaresan (2009) proposed the following modified closure for the drag force:

$$\mathbf{f}_{d,i} = V_{p,i}\left[\beta_{p,i}(\mathbf{u}_i - \mathbf{v}_i) + \sum_{i \neq j} \beta_{p,\mathrm{FMPP},ij}\left(\mathbf{v}_j - \mathbf{v}_i\right)\right]. \tag{42}$$

Here, $\beta_{p,\mathrm{FMPP},ij}$ is a "fluid-mediated particle–particle" friction coefficient that quantifies the effect of velocity differences between individual particle species. It is important to note that this type of drag modification assumes that particles of the same species move with identical speed. Thus, the "species-based" granular temperature is zero (this is important, since it differentiates the fluid-mediated particle–particle drag from the effects that are discussed in Section 5.2). Sundaresan and coworkers later showed that the above idea is also valid for nonzero Reynolds numbers (Holloway et al., 2010). In the same manner as for fixed bed drag closures, the above fluid-mediated force closure has been investigated in the context of EE models. Hence, it is only able to predict the *ensemble-average* drag force acting on particles—*not* per-particle values for the drag.

5.2 Effects Due to Particle Agitation

Along a similar line of thoughts as the discussion in Section 5.1, one can argue that the (average) drag force should be some function of the particle agitation, i.e., the particles' granular temperature. Such an idea was followed recently by two groups: (i) the group around Kuipers suggested adding a "particle mobility" contribution to the drag force coefficient experienced by a static array of particles (Tang et al., 2016). This mobility contribution is linearly dependent on a granular temperature-based Reynolds number. Hence, it is most relevant for intermediate to high Reynolds number flows and always leads to an increase of the drag (compared to that of a fixed bed) in case particles are agitated. (ii) The group around Sundaresan has also investigated the effect of the mobility of particles, but quantified it with a Stokes number (Rubinstein et al., 2016). This group explored rather low Reynolds numbers flows over a large range of Stokes numbers and particle concentrations. Interestingly, this group finds a lower drag for lower Stokes numbers, i.e., a negative correction to the fixed bed drag in case particles are mobile. An opposite trend was predicted by the Kuipers group (Tang et al., 2016).

5.3 Cross Effects

Another fact that is often dropped in suspension flow simulations are cross effects: such effects occur in case of extreme heat transfer rates from the particle surface, since then the particle-near fluid may experience significantly higher temperatures (Ganguli and Lele, 2016). This leads to effects caused by buoyancy and the change of physical parameters (e.g., the gas viscosity). Since rigorous closure models are often not available, a carefully conducted preliminary analysis using dimensionless parameters (e.g., the Grashof number) is helpful in such cases. The interested reader is referred to the Appendix of Tenneti et al. for a more detailed discussion of such effects (Tenneti et al., 2013).

In summary, the closure of Beetstra et al. (2007) is still frequently used, often without correction for polydispersity, the particles' relative speed, or cross effects. Moreover, the above drag closures can only predict the average force acting on an ensemble of particles, but not an individual particle. All these limitations should be taken into account when applied to EE or PU-EL models once per-particle drag correlations are established. The next section illustrates a possible pathway to achieve such an improvement.

5.4 Variability of the Exchange Coefficients

There has been a recent interest to quantify the variability of the per-particle drag coefficient in dense suspensions with the ultimate goal to establish corresponding closures. An early example of this trend is the work of the Kuipers group (Kriebitzsch et al., 2013). More recently, Deen and Kuipers made an effort to characterize the variability of mass transfer coefficients as well (Deen and Kuipers, 2014b). As outlined in Section 4.2.3, it is of paramount importance to define which average fluid quantity was used to calculate exchange coefficients. In most of the previous works, such quantity was obviously motivated by simple Euler–Euler models: for example, Deen and Kuipers use a slice-based cup-mixing temperature (which would be available in a one-dimensional version of an EE model) to calculate the distribution of mass transfer coefficients (Deen and Kuipers, 2014b). It is not surprising that such a choice leads to a very broad distribution of the heat transfer coefficient, and little sense could be made of such data in the past.

What follows is a summary of our recent activities (Municchi and Radl, 2017) to quantify the variability of exchange coefficients in mono- and bidisperse suspensions using the filtering approach.

5.4.1 Drag

With respect to the drag, it is instructive to consider the distribution of a dimensionless per-particle drag force, as well as the trend of the corresponding scaled standard deviation as a function of the average drag on the particles. Such data is shown in Fig. 14 and reveal—surprisingly—no significant

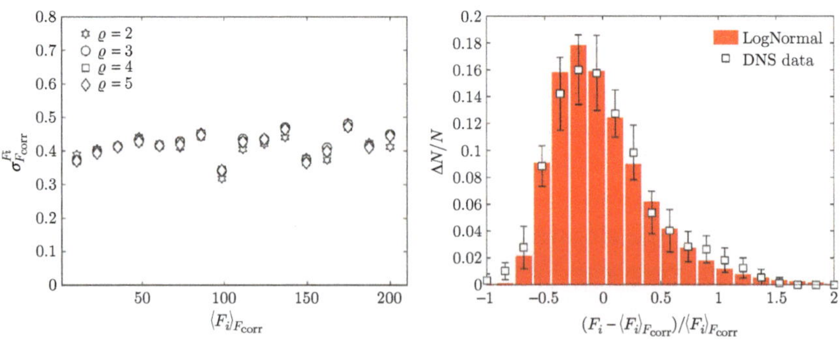

Fig. 14 Scaled standard deviation of the per-particle drag force (*left panel*), as well as number-based distribution of the relative deviation of the conditionally averaged drag force from its ensemble average (*right panel*; the *error bar length* corresponds to two standard deviations of the samples in each bin). ϱ is the dimensionless filter size (Municchi and Radl, 2017).

dependence of the relative standard deviation from the mean (ensemble-average) drag coefficient $\langle F_i \rangle$.

This suggests that the relative deviation of F_i from $\langle F_i \rangle$ can be assumed to be approximately constant. Specifically, values for the standard deviation ranged from 32% to 52% which is generally attributed to the presence of preferred pathways for fluid flow through the particle array. We note in passing that this is in line with previous results in the field (Kriebitzsch et al., 2013) as recently discussed in our previous work (Municchi and Radl, 2017).

Also, the lack of a functional dependence on the mean drag coefficient results in a similar distribution for the relative deviations at different average drag coefficients $\langle F_i \rangle$. This suggests representing the stochastic fluctuation of the drag force using a modified log-normal distribution. This is in contrast to previous work that considered a Gaussian distribution (Kriebitzsch et al., 2013). Our choice of a log-normal distribution is motivated by the following facts:

i. A log-normal distribution is defined in the range $[0,\infty]$ and, thus, naturally predicts a minimum allowed value for the drag coefficient.

ii. As can be seen from Fig. 14, the distribution of relative deviations is skewed.

iii. The chosen log-normal distribution can be defined using only one parameter (i.e., the standard deviation), and it was demonstrated that this parameter is approximately constant within the range of parameters used in our study. In fact, physically, the minimum allowable value for the relative deviation of -1 (i.e., zero drag force) and the expectation value is, by definition, equal to zero. This results in a standard log-normal distribution with a mean value equal to 1.

The choice of a log-normal distribution also allows the stochastic drag coefficient model to be efficiently implemented in PU-EL codes: if ε_i is a randomly generated number between 0 and 1, the value of the fluctuating component of the per-particle drag coefficient can be calculated via an inverse error function that can be tabulated (β is the lower bound of the function and equals -1; α_F is discussed below):

$$F_{i,\sigma} = \beta + \exp\left[\operatorname{erf}^{-1}(2\varepsilon_i - 1)\sqrt{2}\alpha_F\right] \tag{43}$$

In case the coefficients are assumed to be constant (as in our study) the whole expression above can be tabulated before starting the computation. While the above equation can be used to compute the per-particle relative deviations from $\langle F_i \rangle$, the important question remains on how this random deviation should be evolved in time. For the time being we note that our model would allow assigning a specific value for the drag coefficient fluctuation to each particle, and keeping this value throughout a PU-EL simulation.

Finally, the dimensionless drag force acting on a particle i can be computed from:

$$F_i = F_{i,\mathrm{corr}}(1 + F_{i,\sigma}) \tag{44}$$

However, it should be noted that in the limit of infinite dilution and zero Reynolds number, the above stated simple model for the drag force fluctuation (i.e., a constant value) is predicting a drag force fluctuation between -1 and 2. Clearly, within this limit, the resulting drag should return Stokes drag law (i.e., $F_i = 1$). A model for the drag force fluctuation—similar to that shown above—is able to consistently take this limit into account only when one ensures that a suitable expression for the standard deviation α_F is provided. If all the particles tend to experience the same drag force, the modified log-normal distribution must tend to a Dirac delta function, i.e., $\alpha_F = 0$. Future exploration of such a range of parameters is required in order to obtain a suitable expression for α_F.

Another approach that is less consistent, but perhaps more practical and applicable, consists in calculating the particle-based drag force using the following expression:

$$F_i = 1 + \gamma_F F_{i,\mathrm{corr}}\left(1 - \frac{1}{F_{i,\mathrm{corr}}} + F_{i,\sigma}\right) \tag{45}$$

Where a simple functional form for the unknown function γ_F would be the relation:

$$\gamma_F = \frac{F_{i,\mathrm{corr}} - 1}{F_{i,\mathrm{corr}}} \tag{46}$$

It is clear that using the above definition for γ_F enforces the Stokes limit, while it approaches the original closure (i.e., $F_i = F_{i,\mathrm{corr}}(1 + F_{i,\sigma})$) for large values of $F_{i,\mathrm{corr}}$. Since only relatively large Reynolds numbers were probed so far by us, future work is needed to justify the above presented drag force model for drag coefficients close to unity.

5.4.2 Heat and Mass Transfer

Examining the distribution of the per-particle Nusselt number reveals an interesting finding: again one can observe that the scaled standard deviation is approximately constant for the Nusselt number (see Fig. 15), and the

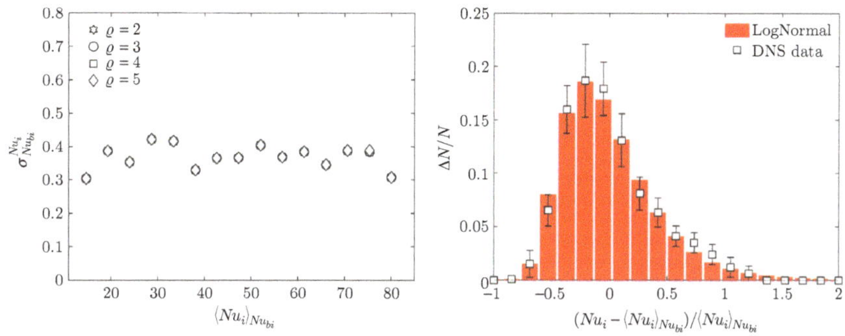

Fig. 15 Scaled standard deviation of the per-particle Nusselt number (*left panel*), as well as number-based distribution of the relative deviation of the conditionally averaged Nusselt number from its ensemble average (*right panel*; the *error bar length* corresponds to two standard deviations of the samples in each bin). ϱ is the dimensionless filter size (Municchi and Radl, 2017).

standard deviation ranges from approximately 30%–45%. Following the same approach as for the drag, one can propose a model for the per-particle Nusselt number based on a log-normally distributed variable (for the fluctuations of the Nusselt number refer to Fig. 15).

Compared to the drag coefficient, the log-normal distribution for the Nusselt number is more peaked (i.e., the standard deviation is smaller), and therefore necessitates the use of an α_{Nu} parameter instead of the corresponding parameter for the drag. In order to represent our PR-DNS data reasonably well, a value of $\alpha_{Nu} = 0.336$ is recommended.

On a final note, one should highlight the importance of ensuring that the predicted Nusselt number approaches the correct values in the limit of infinite dilution and creeping flow. Ideas on how to ensure this have been provided in a recent publication (Municchi and Radl, 2017).

5.5 Wall Effects

Confined suspension flows are a topic of active research since they are of use in a wide range of industrial processes like energy storage, heterogeneous catalytic reactors, or fluid–particle separation processes conducted in microchannels. Also, one might speculate that confinement effects in suspension flows become even more important in the near future: for example, 3D printing technology aiming on producing materials capable to be used at high temperatures (e.g., metals or ceramics) is already reality. This enables the use of complex geometries with characteristic dimensions closer to that of the suspended particles. In such systems wall effects will play a central role.

In addition, the accurate modelling of momentum, heat, and mass transport in dense gas–particle systems is of pivotal importance for designing chemical reactors (Dixon, 2017).

Fig. 16 illustrates a typical temperature distribution in a confined packed bed. It is clearly visible that the fluid velocity is strongly nonuniformly distributed at the top of the particle bed, leading to the formation of jets that originate from the gaps between particles. Consequently, one might question the validity of established correlations for both drag and heat transfer rates in case

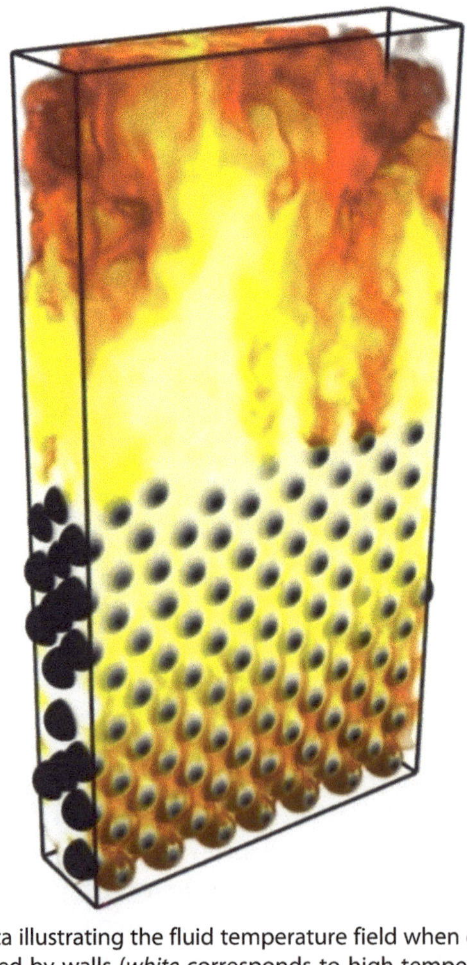

Fig. 16 PR-DNS data illustrating the fluid temperature field when cooling a bed of particles that is confined by walls (*white* corresponds to high temperature, *dark colors* to low temperature).

particles are close to walls. Also, it is clear that suspended particles that are much smaller than that constituting the bed will be transported more quickly in the wall near region due to jet formation. Hence, one could speculate that wall effects are especially important in case of polydisperse suspensions.

A first step to systematically investigate these aspects would be to quantify wall effects in an isolated fashion, i.e., separate them from the curvature effect that is typically included in the analysis (de Klerk, 2003; Mueller, 1992; Theuerkauf et al., 2006; van Antwerpen et al., 2010). Also, little is known for more dilute and moderately dense suspensions, since most previous work explored packed beds only. Considering a wider parameter space is, however, essential when building a robust, generally applicable simulation model. In what follows a brief summary of our recent attempts to perform such a systematic study is presented. The focus was hereby to answer the fundamental question on how one should correct transfer coefficients in PU-EL-based simulations that involve walls. This was achieved by running a large set of PR-DNS. A snapshot of a corresponding simulation result for the temperature distribution is shown in Fig. 17. For details related to this analysis, the interested reader is referred to our latest publication (Municchi and Radl, 2018).

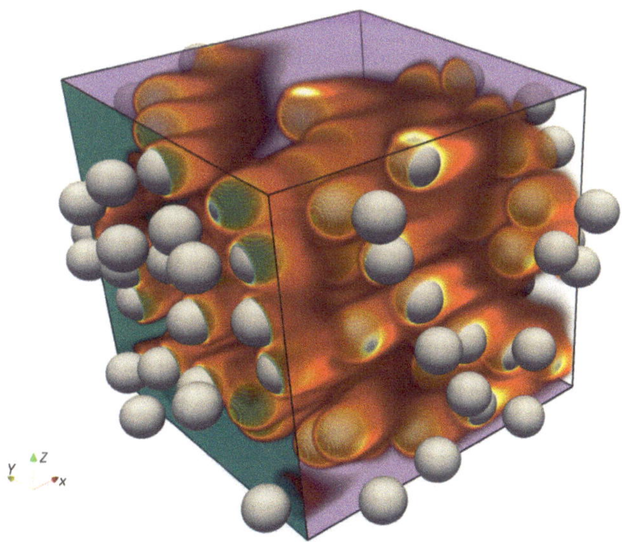

Fig. 17 Typical result for the temperature distribution in a confined gas–particle suspension (particles are confined by walls in the *z*-direction; the other directions are periodic; the particles do not move relative to the adiabatic walls; particles are kept at a fixed temperature).

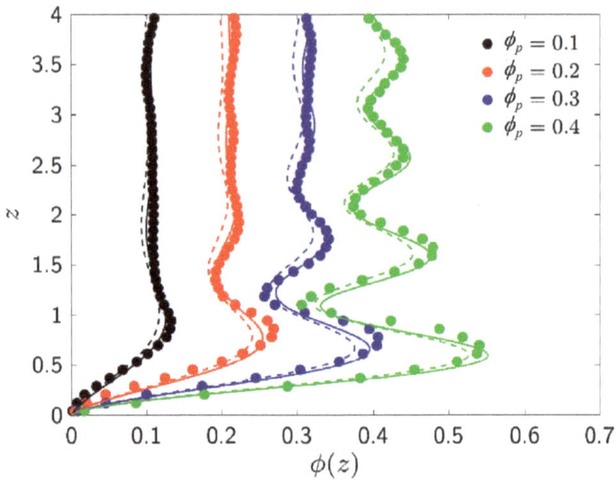

Fig. 18 Average particle volume fraction as a function of the wall normal distance for various values of the average particle volume fraction in a confined bed of otherwise randomly arranged particles (*dots* indicate simulation data, *continuous lines* use the bulk voidage in the model equation, whereas *dashed lines* use the average voidage; Municchi and Radl, 2018).

5.5.1 Particle Concentration and Flow Profiles

Fig. 18 shows the wall normal profile of the particle volume fraction for different values of ϕ_p, which represents the average particle volume fraction in the domain. In current literature, these profiles are often called *wall radial distribution function* due to the cylindrical geometries employed in their evaluation. In what follows it is useful to consider these profiles as perturbations with respect to the homogeneous (bulk) region. From Fig. 18 one can clearly see that both (i) the absolute amplitude and (ii) the wave number of the perturbation are larger for larger values of ϕ_p. This can be related to the higher degree of order that the particles experience near the wall at higher packing fractions.

This wall-induced perturbation of the particle volume fraction has been studied in literature for the densely packed spheres in cylindrical containers. Several correlations were proposed which are all limited to cylindrical coordinates and high packing fractions. Motivated by the previous considerations regarding the behavior of the voidage (Mueller, 1992), the following functional form that relaxes this constraint is proposed:

$$\phi(z) = \phi_b\{1 - J_0(\kappa_1 z)\exp[-\lambda_1 z]\} \tag{47}$$

$$\kappa_1 = 0.875 + 8.55\sqrt{\phi_b} \tag{48}$$

$$\lambda_1 = 0.461 + 0.409\sin(7.42\phi_b + 0.327) \tag{49}$$

Here ϕ_b is the particle volume fraction far away from the wall. This value is unknown in case of a finite domain size, but can be related to the average particle volume fraction (Municchi and Radl, 2018).

The perturbation to the particle volume fraction shown in Fig. 18 gives rise to statistically inhomogeneous velocity (and temperature) fields. Thus, these flow variables are a function of the wall normal distance. An illustration of this behavior is shown in Fig. 19. An especially interesting trend is observed for the mean temperature: it appears that a layer with almost constant temperature gradient normal to the wall is formed. This "boundary layer" is only weakly pronounced for dilute systems (i.e., $\phi_p=0.1$). This is not directly observable from Fig. 19 since the temperature profile is scaled with the bulk temperature. The latter is strongly decreasing for decreasing particle concentrations. Hence, considering the original temperature profiles, one would observe a significantly larger gradient for larger values of ϕ_p compared to $\phi_p=0.1$.

5.5.2 Drag

In order to quantify the effect of wall boundaries on the drag coefficient and Nusselt number, relevant statistical data was collected using an approach similar to that used in Section 5.4. However, this time all data was binned with respect to the particles' wall normal distance, necessitating a much larger amount of raw data to reduce noise in the calculated averages: For each value of the mean particle concentration, data from simulations involving approximately 4000 particles was used. As a key result, it was observed that (as for the wall normal velocity and temperature profiles) the dependence on the

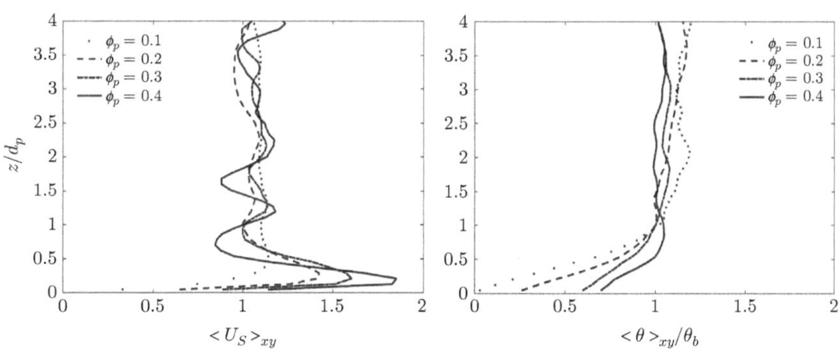

Fig. 19 Wall normal profiles of the average fluid speed (*left panel*; the speed is normalized with the global mean speed in the simulation domain), as well as the average fluid temperature (*right panel*; the temperature is normalized with the fluid's bulk temperature) for a Reynolds number of 400 (Municchi and Radl, 2018).

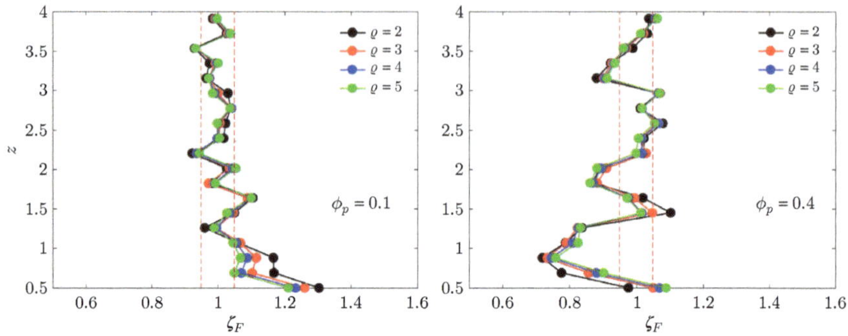

Fig. 20 Wall correction to the drag coefficient as a function of the filter size ϱ and the average particle volume fraction ϕ_p. *Red dashed lines* represent deviations of $\pm 5\%$ from the bulk value.

Reynolds number can be easily dropped once the drag coefficient and the Nusselt number are related to a typical bulk value (this is true for the investigated range of Reynolds numbers, i.e., 100–400). Hence, a master curve for the drag correction can be constructed, which is shown in Fig. 20. Both wall normal profiles have a maximum corresponding to the first particle layer (i.e., $z = 0.5$). A minimum develops when increasing the average particle concentration ϕ_p. It can also be seen that the intensity of the perturbations becomes larger when increasing ϕ_p, which is related to the stronger perturbation in the particle concentration profile. No significant effect of the filter size ϱ was observed, hinting to the fact that filtered fluid velocities are almost independent of ϱ.

5.5.3 Heat and Mass Transfer

Results for the wall correction to the Nusselt number are plotted in Fig. 21. Our results indicate that Nusselt number corrections are in general stronger as those for the drag. This can be tentatively explained by the combined effect of flow (i.e., the velocity gradient) and the abnormal temperature distribution experienced by wall-near particles (see Fig. 19, right panel).

Furthermore, the filter size ϱ has a significant effect on the Nusselt number correction. This effect is extreme for particles closest to the wall, i.e., that located at $z = 0.5$. The enormous sensitivity to ϱ observed for the corrections to the Nusselt number is in contrast to the correction for the drag coefficient. The latter are almost insensitive to the filter size. One could speculate that the origin of this difference is again the abnormal temperature distribution in the vicinity of the wall: Depending on the filter size, a different *filtered*

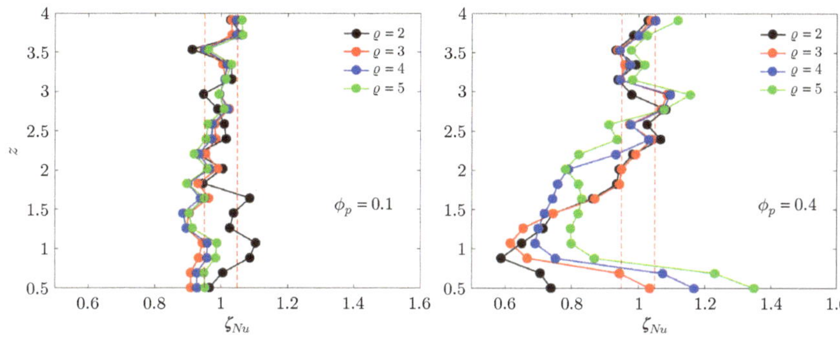

Fig. 21 Wall correction to the Nusselt number as a function of the filter size ϱ and the average particle volume fraction ϕ_p. *Red dashed lines* represent deviations of ±5% from the bulk value.

temperature is experienced by individual particles. Specifically, smaller filter sizes ϱ will lead to lower values for the average fluid temperate as can be anticipated from Fig. 19, right panel. This will lead to a larger difference between the particle and fluid temperature, i.e., the driving average temperature difference. In simple words, a smaller filter size (or a smaller grid size) results in wall-near particles experiencing a different fluid temperature in PU-EL models. Thus, for a given fluid–particle heat flux, the Nusselt number must decrease for a decrease in the filter size in case particles are located in the "temperature boundary layer" observable in Fig. 19. Also, this is exactly what was observed for the correction of the Nusselt number (see our data for $\phi_p = 0.4$ in Fig. 21). In case particles are outside of the "temperature boundary layer," or this layer is only weakly pronounced, the above argument on the filter size effect is expected to break down (see data for $\phi_p = 0.1$ in Fig. 21).

6. RECENT DEVELOPMENTS FOR CLOSURES DESCRIBING MESOSCALE PHENOMENA

Simulations of the mesoscale models discussed in Section 5 provide data that is useful to understand flow phenomena on intermediate scales. In what follows, some developments in this fields are summarized that aim on the quantification of such phenomena in order to provide closure information for the next level on the modelling hierarchy, i.e., macroscale models. These developments are documented by four examples that are thought to best illustrate recent trends, and that were triggered by advances in the field of mesoscale simulations.

6.1 Sequential Fluid and Particle Coarsening

Model coarsening (i.e., establishing closures for macroscale models based on mesoscale simulation) in the context of fluid–particle suspensions has traditionally aimed on using EE models (Holloway, 2012). The key motivation behind this is twofold: first, EE models require less computational resources to support closure development with data. Second, latest kinetic theory-based EE models can be considered equivalent to EL models, at least with respect to picturing relevant clustering phenomena in noncohesive monodisperse fluid–particle suspensions (Fullmer and Hrenya, 2016). Especially, kinetic theory-based rheological models are well developed and are—at the moment—clearly superior to parcel-based strategies (i.e., EL models that do not track individual particles). The latter are often based on heuristic arguments (Snider, 2001).

Still, coarsening of EL model offer advantages over an EE-based strategy:

(i) EL-based models can be used to consider wider spectra of physical models, e.g., cohesive interparticle forces, the fluctuation of the per-particle drag force (Kriebitzsch et al., 2013; Municchi and Radl, 2017), or nonspherical particles. This increases the applicability of model coarsening strategies, since industrially relevant systems may be severely affected by cohesive forces or particle shape. The recent study of Ozel et al. (2017) illustrates this advantage of EL models when deriving models that describe mesoscale phenomena in an excellent way.

(ii) EL models offer a natural tracking of per-particle quantities, e.g., the particle's composition, density, liquid content, or intraparticulate concentration and temperature profiles. This is of immediate relevance for applications in the field of reactive suspensions.

(iii) Coarsening can be performed *sequentially* for the fluid and particle phase when adopting an EL model. As shown in Fig. 22, this offers the key advantage that effects due to fluid and particle coarsening can be separated (Ozel et al., 2016; Radl and Sundaresan, 2014). Such an improved understanding of these two coarsening effects is helpful to improve numerical schemes, or to guide closure development.

When adopting an EL approach for coarsening one arrives formally at an EL-Large Eddy Simulation (LES) approach. Marchioli provides a recent review of EL-LES with a focus on one-way coupled dilute fluid–particle flows (Marchioli, 2017). He correctly states that modelling the fluid velocity seen by the particle is the key challenge for such an approach. Specifically, the challenge is to predict the fluid velocity fluctuation at the particle location, for which a variety of "subgrid-scale" (SGS) closure strategies exist. Unfortunately, Marchioli states that:

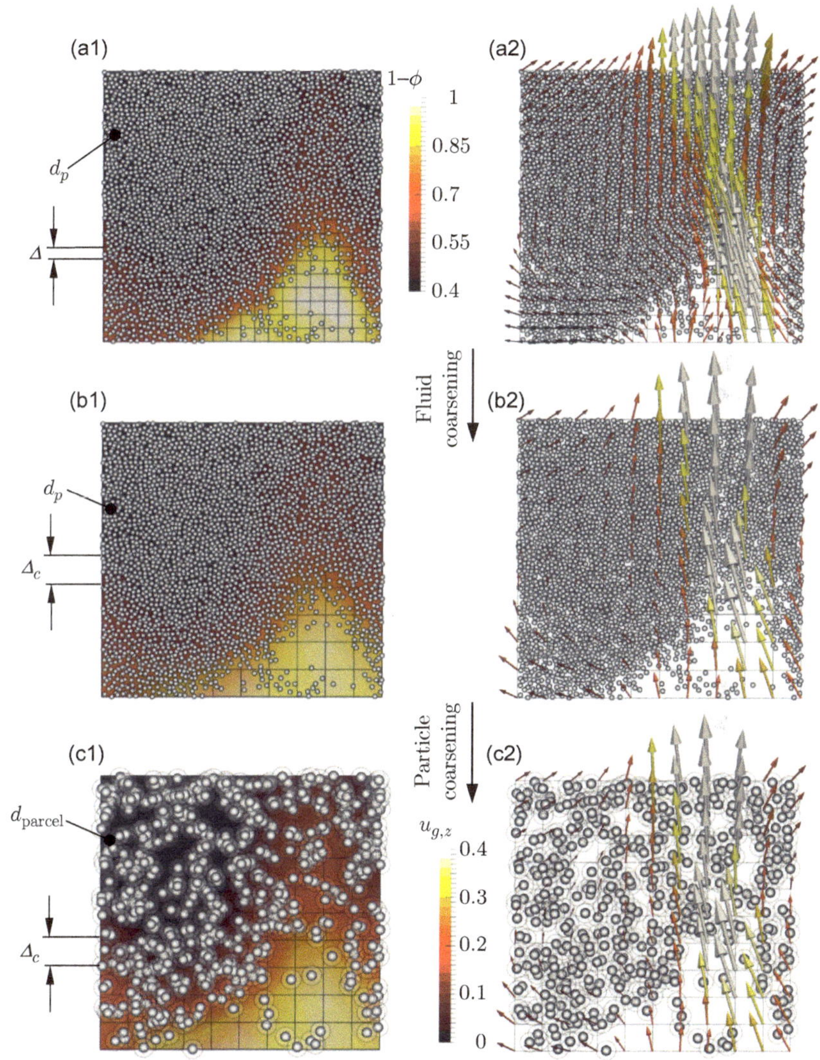

Fig. 22 The concept of sequential fluid- and particle-coarsening in the context of EL models (a1, b1, c1: *colors* indicate the local voidage as per the *top colorbar*; a2, b2, c2: *colors* indicate the vertical fluid speed as per the *bottom colorbar*; d_{parcel} refers to the diameter of the *opaque spheres* in panel c1 and c2; the *semitransparent circles* illustrate the influence region of each parcel, Δ identifies a cell size).

"...no particle SGS models have been developed for semi-dilute or dense suspensions, which would require specific source terms to account for two-way coupling (inter-phase momentum exchange) or four-way coupling (collisions) in the governing equations."

Marchioli, 2017

Thus, one cannot adopt EL-LES closure approaches for dense suspensions in a straightforward manner. Instead, one is—more or less—forced to rely on a somewhat more pragmatic approach that has been outlined in Section 2.6.2.

6.2 Non-Boussinesq Stress Models

Current models describe mesoscale solids stresses—in analogy to the Boussinesq approximation for turbulent stresses—based on a filtered solids pressure and a filtered solids viscosity. This previously adopted approach is therefore heavily reliant on the assumption that mesoscale velocity fluctuations of the solid phase are isotropic. However, latest developments in the field of multiphase Reynolds stress models (RSMs)—even for comparably dilute particle laden flows—indicate that mesoscale stresses are highly anisotropic (Capecelatro et al., 2016).

The group around Amini (SINTEF, Norway) has attacked the associated question whether stress models that are beyond the Boussinesq assumption can be built. While limited to a two-dimensional proof of principle study, the first fruits of this research could be harvested just recently (Cloete et al., under review). One core result is that mesoscale shear stresses do not correlate well with the corresponding components of the filtered shear rate tensor. In other words, the core issue is the assumption inherent to a scalar mesoscale viscosity. Furthermore, it was found that the conventional isotropic mesoscale pressure approach offers a reasonable approximation of the filtered resolved data.

Subsequently, a rigorous marker screening study was performed, and it was found that products of velocity gradients were much better markers compared to the shear rate (Cloete et al., under review). Fig. 23 illustrates this finding, in which the particle-phase velocity gradient product marker was calculated via:

$$M_{xy} = \overline{\alpha_s}\left(\frac{d\overline{\mathbf{v}}_x}{dx}\frac{d\overline{\mathbf{v}}_y}{dx} + \frac{d\overline{\mathbf{v}}_x}{dy}\frac{d\overline{\mathbf{v}}_y}{dy}\right) \tag{50}$$

This gives new hope that closures for each component of the mesoscale stress tensor can be built with relative ease. Hence, these closures may rely on tensor-valued markers.

6.3 Closures for Cohesive Systems

Cohesive forces play a central role in many industrial particle processes, e.g., fluidization of fine dry powders, or wet fluidization (e.g., wet granulation, fluidized bed drying, fluid coking, etc.). Taking the example of

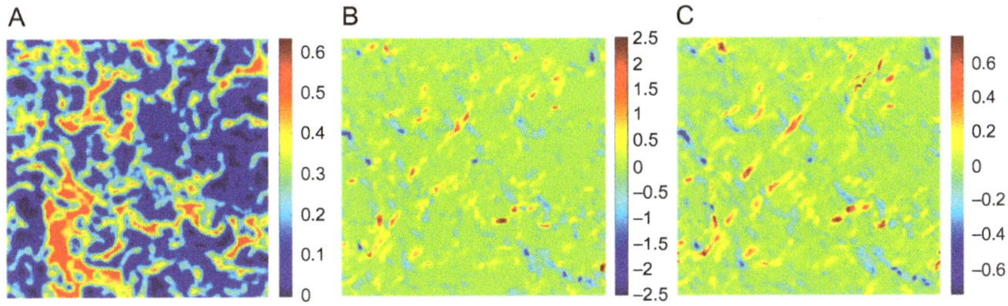

Fig. 23 Comparison of the particle-phase mesoscale shear stress to its corresponding gradient-based marker for a domain-averaged solids volume fraction of 0.2 and a dimensionless filter size of 3.47. (A) The filtered solids volume fraction, (B) filtered shear stress $\hat{\boldsymbol{\Sigma}}_{\boldsymbol{s,fil,xy}}$, (C) gradient-based marker $\boldsymbol{M_{xy}}$. *Figure by courtesy of Schalk and Jan Hendrik Cloete, as well as Shahriar Amini, SINTEF, Norway.*

dry powder fluidization, and focusing on the formation of bubbles, a number of breakthrough computational studies have been published recently. Building on pioneering work in the field (Rhodes et al., 2001; Ye et al., 2004), two recent studies stand out: (i) the work of Galvin and Benyahia showed how particle size distributions and cohesion affect fluidization properties (Galvin and Benyahia, 2014). (ii) Sundaresan and coworkers (Gu et al., 2016a, b) found that even two particle size classes and cohesion are enough to predict the fluidization regime of Geldart A particles. These previous studies were descriptive and supported findings on bubble sizes. While this is not of immediate relevance to improve closures, they contributed to speeding up PU-EL-based simulations by proposing stiffness correction rules. Such rules are of large interest in the field (Hærvig et al., 2017).

Recent developments in the field of wet fluidization were—*inter alia*—based on a more profound understanding of liquid bridge formation dynamics (Wu et al., 2016, 2017). This improved understanding of how liquid bridges form allows to map the physical properties of a wet fluidized bed to its expected fluidization state (Boyce et al., 2017; Wu et al., 2018). Also, a first step toward using simulation data to develop a quantitative description of mesoscale phenomena (i.e., agglomerate formation, and the associated increase in the fluid–particle slip velocity) was already taken (Girardi et al., 2016). More recent studies (Wu et al., 2018) confirm these results of Sundaresan and coworkers. Future work can now expand the ideas that have been introduced with respect to coarsening EL models for wet granular systems via filtering. Thus, there is hope to arrive at macroscale models of wet fluidized systems that are of immediate relevance for industry.

Finally, a note must be made related to cohesive forces developing in electrostatically charged gas–particle suspensions: The complexity inherent to tribocharging has long prevented accounting for electrostatic forces when simulating such systems. Fortunately, some key challenges have been overcome just recently (Kolehmainen et al., 2016, 2017a; Shinbrot et al., 2017). A central result from these studies is that there is now an improved understanding how simulations should be performed, e.g., with respect to stiffness scaling (Kolehmainen et al., 2017b). Still, closures for predicting drag forces or stresses useful for macroscale models have not been distilled so far. Such models are clearly needed as summarized in a recent review (Fotovat et al., 2017). That the development of such mesoscale models for electrostatically charged fluid–particle systems is hindered by the enormous computational cost (of EL models) is clear: as of now, the typical amount of particles that can be studied with high fidelity tribocharging models is in the order of 50,000 (Fotovat et al., 2017). Unless rigorous EE models for modelling cohesion and charge uptake in tribocharged fluid–particle systems are fully developed, this limitation will persist. Fortunately, first steps in developing such EE models have been made just recently (Kolehmainen et al., 2017c).

6.4 Closures Based on a Deconvolution Approach

The concept of deconvolution analysis has been widely exploited in the field of single-phase turbulent flow as a promising approach for so-called structural closures in the context of LESs (Sagaut, 2006). This type of closures relies on no a priori knowledge of the flow phenomena described, but is solely based on a formal description of the spatial filtering procedure. The main idea of deconvolution analysis is the inversion of the filtering procedure, i.e., to employ a "defiltering" procedure to obtain unfiltered quantities (e.g., mesoscale stresses) from filtered data. The latter is provided by a macroscale model, and hence available at a comparably coarse computational grid resolution level. "Defiltered" quantities can then be used in a straightforward manner to describe the effect of the phenomenon of interest (e.g., mesoscale stresses caused by particle clustering). Consequently, this information can be used to close the macroscale model equations. Closures based on a deconvolution analysis are typically referred to as "approximate deconvolution models" (ADM), a wording that might be misleading since no (physical) model equations are actually solved.

ADM has been introduced to the field of gas–particle flows already in 2006, however, only in the context of relatively dilute flows (Kuerten, 2006). For dense flows, an attempt to test the virtues and merits of ADM was made only recently (Schneiderbauer and Saeedipour, 2018). Specifically, deconvolution was applied to analyze the effect of clustering on the interphase momentum exchange rate, as well as on the mesoscale solids stress in vertical channel flow. It was shown that ADM is indeed a promising alternative to strategies that rely on algebraic closures (e.g., those discussed in Section 6.2). However, this was demonstrated only for small spatial extensions of the filter size, corresponding to comparably fine computational grids to be used when simulating a macroscale model.

Despite the current limitations of ADM-based closures, Schneiderbauer and Saeedipour highlighted two key advantages of such an approach when applied to dense fluid–particle flows (Schneiderbauer and Saeedipour, 2018):

i. The anisotropy of mesoscale stresses is correctly predicted by the ADM-based closure for the situation of a vertical channel flow. This gives hope that solids stresses can be reliably predicted in flow situations involving walls.

ii. A number of assumptions when closing the filtered drag force can be dropped, most important that of assuming that the mesoscale correction factors for the drag force are scalar valued.

The application of the deconvolution approach may certainly improve the quantitative understanding of mesoscale phenomena, and hence the quality of closures in the field of macroscale models. Unfortunately, for dense gas–particle flows encountered in large-scale engineering applications (e.g., fluidized beds), ADM appears to be less useful at the moment. This is simply due to the (comparably) fine computational grids that need to be used as already discussed earlier, and due to the lack of benchmark simulations for such applications. Also, factors like (i) the numerically efficiency, and hence the affordability when using deconvolution in the context of macroscale models, or (ii) the ability to include more physics (e.g., polydispersity, species transport, chemical reactions) are currently unexplored and may hinder the widespread use of ADM.

7. CONCLUSION AND OUTLOOK

The application of data filtering, i.e., spatial averaging, together with advances in computational tools for predicting dense gas–particle flows in high detail, has equipped researchers with powerful tools to investigate these systems. Combined with a thorough theoretical understanding, these tools have been used to develop new closure to be used in meso- and macroscale models. Mostly, this was limited to monodisperse and noncohesive systems made of spherical particles. Since simulators are increasingly based on the

Euler–Lagrange approach, these limitations can be—hopefully—removed in the near future with relative ease. Certainly, success along this way of thoughts depends on the application. It is certain that cohesive forces caused by electrostatics, as well as polydisperse systems with a broad size distribution offer some of the most significant challenges from the point of computations.

There is a plurality of possible research directions that one could follow, and that rely on spatial averaging as their commonality. Fig. 24 provides an overview of such research areas in which one could branch out. Clearly, the development of closures is certainly the broadest field of application of the filtering approach. However, in parallel it will be necessary to update and upgrade existing models to fully exploit new closures: An example is the new generation of tensor-valued closure laws (e.g., for the mesoscale stress as mentioned in Section 6.2) to be used in macroscale models. Without a significant upgrade of the corresponding tools that simulate these macroscale models, it will not be possible to exploit the predictive capabilities of such new closures.

In what follows, three research directions for which clear research tasks can be defined are summarized in greater detail.

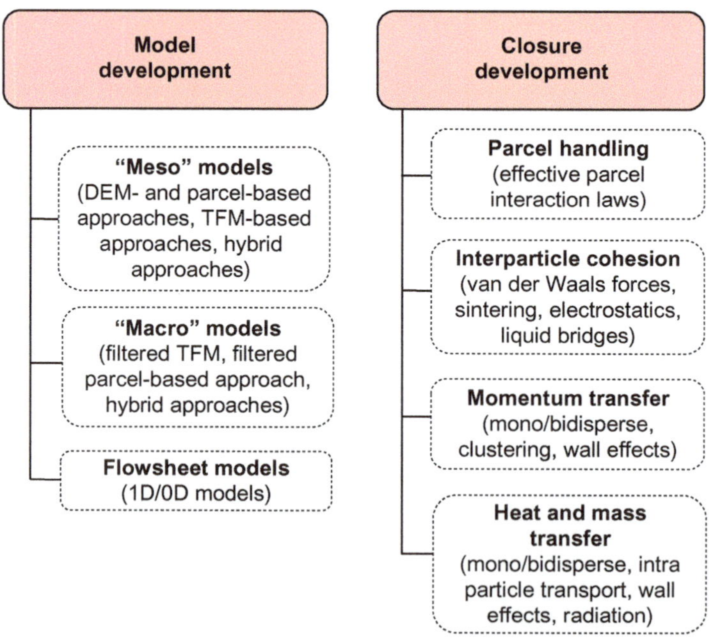

Fig. 24 Overview of possible future research areas in the field of fluid–particle suspensions that benefit from the development of tools for efficient spatial data filtering.

7.1 Research Direction 1: Advanced Euler–Lagrange Models

First and foremost, it would be wise to combine findings between different studies related to fluid–particle drag forces. A prime example are the recent studies of the Kuipers group (Tang et al., 2016) and the Sundaresan group (Rubinstein et al., 2016) on the effect of particle agitation on the drag coefficient. While on a first view, it appears that these studies contradict each other, they most likely probed two different extremes in the large parameter space spanned by sedimenting fluid–particle suspensions. Looking into statistics obtained from spatial data filtering could help to shed light into this possible discrepancy. Some pioneering work has been made that considers the effect of the suspension's microstructure on the drag (Akiki et al., 2017; Holloway et al., 2012), which could also help to aid in unifying this fuzzy picture. Certainly, such ideas should be extended in the close future.

Second, significant attempts were already made to exploit the key advantages of EL models. These are (i) the ability to study cohesive particulate systems (Girardi et al., 2012, 2016; Greidinger et al., 2014; Ozel et al., 2017), (ii) the natural handling of per-particle and per-contact properties (e.g., size, density, liquid content, liquid bridge volume, (Wu et al., 2014, 2016), or particle temperature), (iii) the ability to resolve intraparticle property profiles (Forgber et al., 2017), (iv) the direct handling of wall collisions, and (v) the direct description of flows involving nonspherical particles.

Macroscale simulations that rely on a verified and validated Euler–Lagrange model would open new research directions, especially for cohesive and polydisperse gas–particle suspensions. Building confidence in such models should mainly focus on improving closure for the drag and the particle-phase stress. Specifically, more robust closures must be build that are applicable to situations involving walls, liquid–particle suspensions, or strongly sheared systems.

7.2 Research Direction 2: Broader Application of Filtering Tools

The application of spatial filtering to parameterize closures necessitates the selection of a suitable simulation setup to provide the raw data. So far, mostly periodic domains were chosen. In these systems a nontrivial solution of the governing equations evolves naturally once the initial conditions and the system parameters are appropriately set. For example, in freely sedimenting suspensions such solutions are triggered by an instability mechanism that drives the flow by gravity (Sundaresan, 2003). However, solutions from periodic domains are unsatisfactory to some degree, since it is not known if these solutions are fully representative of real-world applications. Hence, it would be desirable to compare statistics from periodic domains with "real-world" statistics collected when simulating a specific application. Fortunately, some inspiration for such a future extension can be gained from the related field of turbulent combustion as discussed next. In this field DNS are popular and have become a mainstream already decades ago. The simulation setups used in the field of combustion can be readily mapped to other fields, such as mesoscale EL models of dense suspensions. Specifically, Moureau et al. (2011), as well as Domingo and Vervisch (2017) discuss the state of the art in DNS of turbulent combustion. In these applications, three types of DNS setups are used:

- **(i)** DNS of canonical flow situations (e.g., forced synthetic turbulence in a periodic domain, Ten Cate et al., 2004),
- **(ii)** an "embedded DNS box" approach (i.e., a DNS region is directly coupled to a full LES and runs cocurrently, Domingo and Vervisch, 2017), and
- **(iii)** DNS of small domains (this approach is useful for confined particle beds, e.g., see Singhal et al. (2017)).

With respect to (i), the Eswaran–Pope forcing method (Eswaran and Pope, 1988) is typically employed (Ten Cate et al., 2004). This approach appears to be suitable also for anisotropic forcing, which is an essential characteristic of gravity-driven instabilities in fluid–particle suspension. An alternative canonical flow situation is simple (turbulent) shear flow as recently demonstrated by the group around Fox (Baker et al., 2016; Fox et al., 2016). This shear flow setup allows the specification of a mean shear rate as an additional simulation input parameter. Similar pathways could be followed in future, for example, when studying wet fluidized beds. However, one must keep in mind that the parameter space to be explored (e.g., the relative importance of the mean shear rate, or the orientation of gravity relative to mean shear) results in additional challenges with respect to computational effort.

The "embedded DNS box" approach (i.e., item (ii) in the above list) was followed by Pirker et al. (2013) and is also recommended by the recent study of Domingo and Vervisch (2017). This approach was not followed in previous studies that attempted to derive a drag model for clustered gas–particle suspensions. Hence, this approach would offer an interesting alternative to the fully periodic box setup, or the "periodic riser" (Cloete et al., 2017a, under review; Ozel et al., 2013) approach that is well explored in literature.

The filtering approach discussed in this chapter is certainly not limited to bridging a single gap spanned by two characteristic length scales.

Thus, one may want to use filtering progressively, i.e., obtain filtered models from simulations that already employ a filtered model. However, for such a progressive filtering approach the choice of the simulation domain, including appropriate boundary conditions (and perhaps forcing), is not obvious.

7.3 Research Direction 3: Exploration of Intraparticle Transport Phenomena

The new software tool "ParScale" (Radl et al., 2015b) was integrated with the LIGGGHTS® and CFDEM® simulator (results of a typical simulation are shown in Fig. 25). This enhanced simulation platform is now in use to quantify the effect of intraparticle transport phenomena on, e.g., the performance of a heterogeneous reactor. Also, models to predict interparticle heat transfer rates due to radiation are available (Forgber and Radl, 2018). Most important, $O(10^6)$ particles, each discretized by typically 10 intraparticle grid points,

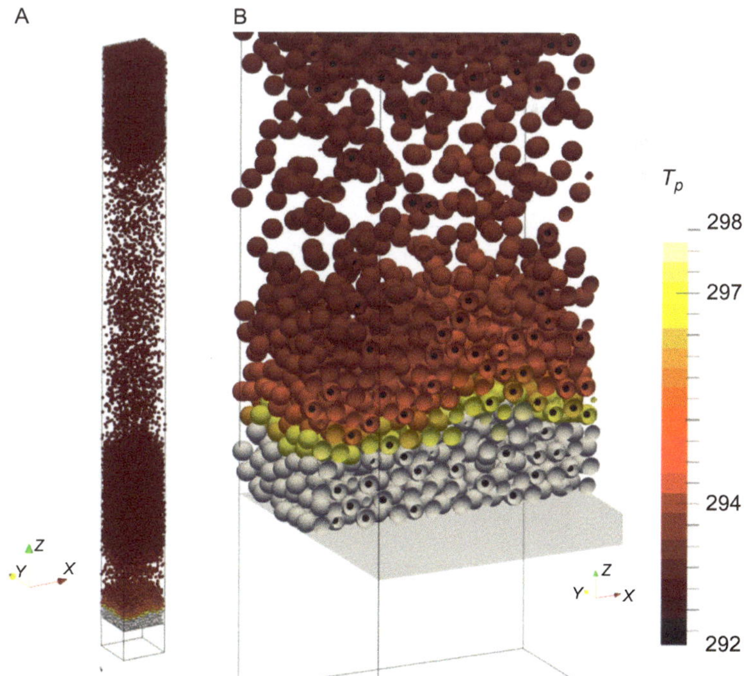

Fig. 25 Particle temperature distribution T_p in a heated fluidized bed (Radl et al., 2015a). (A) full system, (B) close-up view near the fluid inlet; 25,000 particles, each particle is discretized using 10 radial grid points; the characteristic Biot number is 16.3, domain width/depth/height: $15 \cdot d_p/15 \cdot d_p/200 \cdot d_p$, the domain is x- and y-periodic, particle diameter $d_p = 2$ mm, inlet superficial velocity $U_s = 2$ m/s, fluid inlet temperature $T_{in} = 323$ [K].

can be tracked when moving through the simulation domain. The ability to study such a large particle number is achieved by using LIGGGHTS® parallel data containers. These containers solve the issue of efficient data transfer once a particle moves across a processor boundary, and hence a different processor takes over the calculation.

Similar ideas on establishing a similar EL model-based platform were followed by other groups (Oschmann et al., 2016). The latter work used a full (i.e., three-dimensional) discretization of each particle, relying on either a Cartesian or a spherical inner coordinate system. Situations characterized by a small particle Reynolds number (which lead to significant mean temperature gradients in the ambient fluid), as well as situations in which particles are located near walls (Dixon, 2017) may necessitate such a three-dimensional intraparticle discretization. Finally, one must mention the enormous computational expense that is needed to solve EL models with intraparticle discretization. For example, the study of Mahmoudi et al. (2016) considered only a (pseudo) two-dimensional moving bed of particles consisting of a few thousand particles. Clearly, computational efficiency is still a critical bottleneck. It appears that in the coming decade this bottleneck will persist, even though it might be alleviated by the decreasing cost of computing.

ACKNOWLEDGMENTS

Large parts of our studies cited in this chapter were funded by the European Commission through Grant no. 604656 (NanoSim). The authors of this chapter acknowledge the support of NAWI Graz by providing access to VSC-3 and dcluster for carrying out computer simulations. CFDEM® and LIGGGHTS® are trademarks of DCS Computing GmbH. The authors thank Jakob Redlinger-Pohn (TU Graz) for proof-reading the manuscript.

APPENDIX. DERIVATION OF AN EULER–LAGRANGE THEORY OF DISPERSED MULTIPHASE FLOWS

In this Appendix, we provide a “bottom-up” derivation of the governing equations for the fluid phase in the frame of an Euler–Lagrange formulation. Specifically, the continuous phase fields are averaged within a volume V for which no assumption is made with regards to shape or extension. In general $V = V(\boldsymbol{x})$, which defines the spatial coordinate of the averaged fields. In order to simplify the notation, we do not make any distinction between the integration domain and the volume (or mass) enclosed within that domain, i.e., V is referred both to the integration domain and its volume.

This Appendix is structured as follows: in Appendix A.1, we define the basic quantities and operators needed to carry out our analysis. Such operators are applied to the microscopic governing equations in Appendix A.2. Finally, in Appendix A.3 we simplify the EL equations through the use of boundary conditions and further assumptions.

A.1 Definitions and basic results

We define the *mass-based average* (for example, for the velocity field u_i) as:

$$\widehat{u}_i(\boldsymbol{x}, t) = \frac{1}{\mathcal{M}_f(\boldsymbol{x}, t)} \int_{\mathcal{M}_f(\boldsymbol{x}, t)} u_i(\boldsymbol{x}', t) d\mathcal{M}_f \tag{A1}$$

Where $\mathcal{M}_f(\boldsymbol{x}, t)$ is the fluid mass enclosed in the averaging volume V. Notice that this quantity is dependent on space and time because (i) V is a function of the spatial coordinate and (ii) the amount of fluid within V can change due to the movement of particles and variations in the fluid density.

To find a clearer expression for integral Eq. (A1), we introduce the *phase volume average* $V_f(\boldsymbol{x}, t)$, which is the average over the fraction of the averaging volume that is occupied by the fluid and we apply it to the fluid density field $\rho(\boldsymbol{x}, t)$:

$$\widetilde{\rho}(\boldsymbol{x}, t) = \frac{1}{V_f(\boldsymbol{x}, t)} \int_{V_f(\boldsymbol{x}, t)} \rho(\boldsymbol{x}', t) dV_f \tag{A2}$$

Which, considering that $\mathcal{M}_f(\boldsymbol{x}, t) = \widetilde{\rho}(\boldsymbol{x}, t) V_f(\boldsymbol{x}, t)$, results in $d\mathcal{M}_f = \rho(\boldsymbol{x}, t) dV_f$, which allows a change of variables in integral Eq. (A1). Notice that V_f is also a function of space and time, but its time dependence is only due to changes in the dispersion phase, i.e., particles that enter/leave the domain V or that modify their volume (for example, through shrinking). Therefore, we separate the effect of density variability from the pure geometrical contribution to the total mass of fluid in V.

Finally, we introduce the *total volume average*, i.e., over the whole filtering volume $V(\boldsymbol{x})$:

$$\overline{\phi}(\boldsymbol{x}, t) = \frac{1}{V(\boldsymbol{x})} \int_{V(\boldsymbol{x})} \phi(\boldsymbol{x}', t) dV \tag{A3}$$

Where $\phi(\boldsymbol{x}, t)$ is the indicator function, which is 1 when $\mathbf{x}$ lies in the fluid phase and 0 otherwise. Therefore, we can make an additional change of variables, simply by employing $V_f(\boldsymbol{x}, t) = \overline{\phi}(\boldsymbol{x}, t) V(\boldsymbol{x})$, therefore obtaining

$dV_f = \phi(\boldsymbol{x}, t)dV$. Notice how the time dependence of the integration domain has been removed by introducing ϕ.

In the following, we may use a shorthand notation for the *total volume averaging*

$$\overline{\phi}(\boldsymbol{x}, t) = \fint_V \phi(\boldsymbol{x}', t)\, dV \tag{A4}$$

Such notation has the remarkable advantage (over, for example, the classical $\overline{(*)}$) of not hiding the integral nature of the operator while hiding the division by V, which is not important for the final form of the equations.

All the previous definitions can be employed to highlight how different averaging procedures come into play:

$$\begin{aligned}\frac{1}{\mathcal{M}_f(\boldsymbol{x}, t)}\int_{\mathcal{M}_f(\boldsymbol{x}, t)} u_i(\boldsymbol{x}', t)d\mathcal{M}_f &= \frac{1}{V_f(\boldsymbol{x}, t)\widetilde{\rho}(\boldsymbol{x}, t)}\int_{V_f(\boldsymbol{x}, t)} \rho(\boldsymbol{x}', t)u_i(\boldsymbol{x}', t)dV \\ &= \frac{1}{V(\boldsymbol{x})\overline{\phi}(\boldsymbol{x}, t)\widetilde{\rho}(\boldsymbol{x}, t)}\int_{V(\boldsymbol{x})} \phi(\boldsymbol{x}', t)\rho(\boldsymbol{x}', t)u_i(\boldsymbol{x}', t)dV = \frac{\overline{\phi\rho u_i}}{\overline{\phi}\widetilde{\rho}} = \frac{\widetilde{\rho u_i}}{\widetilde{\rho}} = \widehat{u}_i\end{aligned} \tag{A5}$$

In the following, we will often omit the space-time dependence in order to obtain a compact notation.

We introduce the *surface integral average operator* (i.e., average over the immersed surfaces), which is related to the volume average by:

$$\begin{aligned}\fint_V \frac{\partial u_i}{\partial x_i} dV &= \frac{1}{V}\int_V \frac{\partial u_i}{\partial x_i} dV = \sum_{s=0}^{N_s} \frac{1}{V}\int_{S_s} u_i n_{s,i} dS_s = \sum_{s=0}^{N_s} \frac{S_s}{V}\cdot\frac{1}{S_s}\int_{S_s} u_i n_{s,i} dS_s \\ &= \sum_{s=0}^{N_s} \gamma_s \fint_{S_s} u_i n_{s,i} dS_s\end{aligned} \tag{A6}$$

Where S_s is the immersed surface s (N_s is the number of immersed surfaces), $n_{s,i}$ is the i component of the normal to surface s, and γ_s is the *specific surface area* of surface s. Notice that N_s includes the number of particles N_p and the other domain boundaries.

Generally, the indicator function can be casted in the following form:

$$\phi(t, x_1, x_2, x_3) = \sum_{i=1}^{N_p} \mathcal{H}[\Upsilon_{s,i}(t, x_1, x_2, x_3)] \tag{A7}$$

Where N_p is the number of particles and $\mathscr{H}$ is the Heaviside function. The specific shape of an immersed body i is defined by the function $\Upsilon_{s,i}$. Notice that $\Upsilon_{s,\,i}$ is zero at the immersed surface and therefore, by the theorem of implicit functions, the gradient of $\Upsilon_{s,i}$ corresponds to the field of vectors normal to the immersed surface i.

Therefore, the gradient of ϕ can be evaluated:

$$\oint_V \frac{\partial \phi}{\partial x_j} dV = \sum_{i=1}^{N_p} \oint_V \frac{\partial \Upsilon_{s,i}}{\partial x_j} \delta(\Upsilon_{s,i}) dV = \sum_{i=1}^{N_p} \gamma_i \oint_{S_i} n_{s,j} dS_i \tag{A8}$$

Where δ stands for the Dirac delta function.

A.2 Derivation of the volume-averaged governing equations

In the following, we will perform a "step-by-step" derivation of the coarse-grained governing equations by volume averaging.

A.2.1 Microscopic equations

We begin by writing down the microscopic equations for mass (ρ), momentum (ρu_i), and scalar (θ) balance:

$$\frac{\partial \rho}{\partial t} + \frac{\partial}{\partial x_j}(\rho u_j) = 0 \tag{A9}$$

$$\rho \left[\frac{\partial u_i}{\partial t} + \frac{\partial}{\partial x_j}(u_i u_j) \right] = -\frac{\partial p}{\partial x_i} + \frac{\partial \tau_{ij}}{\partial x_j} + \rho g_i \tag{A10}$$

$$\frac{\partial \theta}{\partial t} + \frac{\partial}{\partial x_j}(u_j \theta) = \frac{\partial \Phi_j}{\partial x_j} + \Theta \tag{A11}$$

Where p is the dynamic pressure, g_i is the gravitational acceleration, τ_{ij} is the viscous stress tensor, Φ_j is a scalar flux, and Θ is a scalar source term. We will also assume a general form for the boundary conditions in the form of $a_s\psi(s,t) + b_s\partial\psi/\partial n = c_s$ where a_s, b_s, and c_s are the parameters referring to boundary s and ϕ can be ρ, p, u_i, or θ, being n the normal to the surface. It is easy to show that typically used boundary conditions can be obtained via an appropriate choice of these three parameters.

A.2.2 Coarse-grained mass balance

1. Multiply by indicator function:

$$\phi \frac{\partial \rho}{\partial t} + \phi \frac{\partial}{\partial x_j}(\rho u_j) = 0 \tag{A12}$$

2. Average over the filtering volume:

$$\fint_V \phi \frac{\partial \rho}{\partial t} dV = -\fint_V \phi \frac{\partial}{\partial x_j}(\rho u_j) dV \tag{A13}$$

3. Use integration by parts:

$$\fint_V \left[\frac{\partial}{\partial t}(\phi\rho) - \rho \frac{\partial \phi}{\partial t} \right] dV = -\fint_V \left[\frac{\partial}{\partial x_j}(\phi \rho u) + \rho u_j \frac{\partial \phi}{\partial x_j} \right] dV \tag{A14}$$

4. Use derivative of indicator function and definition of volume surface average for the last RHS term:

$$\fint_V \rho u_j \frac{\partial \phi}{\partial x_j} dV = \sum_{s=0}^{N_p} \gamma_s \fint_{S_s} \rho u_j n_{s,j} dS_s \tag{A15}$$

5. Take derivative out of the integrals:

$$\frac{\partial}{\partial t}(\overline{\phi}\widetilde{\rho}) - \fint_V \rho \frac{\partial \phi}{\partial t} dV = -\frac{\partial}{\partial x_j}(\overline{\phi}\widetilde{\rho}\widehat{u_j}) + \sum_{s=0}^{N_p} \gamma_s \fint_{S_s} \rho u_j n_{s,j} dS_s \tag{A16}$$

6. Rearranging:

$$\frac{\partial}{\partial t}(\overline{\phi}\widetilde{\rho}) + \frac{\partial}{\partial x_j}(\overline{\phi}\widetilde{\rho}\widehat{u_j}) = \fint_V \rho \frac{\partial \phi}{\partial t} dV + \sum_{s=0}^{N_p} \gamma_s \fint_{S_s} \rho u_j n_{s,j} dS_s \tag{A17}$$

A.2.3 Coarse-grained momentum equation

1. Multiply by ϕ:

$$\phi \frac{\partial}{\partial t}(\rho u_i) + \phi \frac{\partial}{\partial x_j}(\rho u_i u_j) = -\phi \frac{\partial p}{\partial x_i} + \phi \frac{\partial \tau_{ij}}{\partial x_j} + \phi \rho g_i \tag{A18}$$

2. Integrate over the filtering volume:

$$\fint_V \phi \frac{\partial}{\partial t}(\rho u_i) dV + \fint_V \phi \frac{\partial}{\partial x_j}(\rho u_i u_j) dV = -\fint_V \phi \frac{\partial p}{\partial x_i} dV + \fint_V \phi \frac{\partial \tau_{ij}}{\partial x_j} dV + \fint_V \phi \rho g_i dV \tag{A19}$$

3. Time derivative

(a) Use integration by parts:

$$\fint_V \phi \frac{\partial}{\partial t}(\rho u_i)\, dV = \fint_V \frac{\partial}{\partial t}(\phi \rho u_i)\, dV - \fint_V \rho u_i \frac{\partial \phi}{\partial t} dV \tag{A20}$$

(b) Exchange integration with derivation:

$$\fint_V \phi \frac{\partial}{\partial t}(\rho u_i)\, dV = \frac{\partial}{\partial t}\left(\overline{\phi}\widetilde{\rho}\,\widehat{u_i}\right) - \fint_V \rho u_i \frac{\partial \phi}{\partial t} dV \tag{A21}$$

4. Nonlinear term

(a) Integration by parts:

$$\fint_V \phi \frac{\partial}{\partial x_j}\left(\rho u_i u_j\right) dV = \fint_V \frac{\partial}{\partial x_j}\left(\phi \rho u_i u_j\right) dV - \fint_V \left(\rho u_i u_j\right) \frac{\partial \phi}{\partial x_j} dV \tag{A22}$$

(b) Use surface integration:

$$\fint_V \phi \frac{\partial}{\partial x_j}\left(\rho u_i u_j\right) dV = \frac{\partial}{\partial x_j}\left(\overline{\phi}\widetilde{\rho}\,\widehat{u_i u_j}\right) - \sum_{s=0}^{N_p} \gamma_s \fint_{S_s} \rho u_i u_j n_{s,j} dS_s \tag{A23}$$

5. Pressure term

(a) Integration by parts:

$$-\fint_V \phi \frac{\partial p}{\partial x_i} dV = -\fint_V \frac{\partial}{\partial x_i}(\phi p)\, dV + \fint_V p \frac{\partial \phi}{\partial x_i} dV \tag{A24}$$

(b) Use phase volume average and derivative of ϕ:

$$-\fint_V \phi \frac{\partial p}{\partial x_i} dV = -\frac{\partial}{\partial x_i}\left(\overline{\phi}\widetilde{p}\right) + \sum_{s=0}^{N_p} \gamma_s \fint_{S_s} p n_{s,i} dS_s \tag{A25}$$

6. Stress tensor

(a) Use integration by parts:

$$\fint_V \phi \frac{\partial \tau_{ij}}{\partial x_j} dV = \fint_V \frac{\partial}{\partial x_j}\left(\phi \tau_{ij}\right) - dV \fint_V \tau_{ij} \frac{\partial \phi}{\partial x_j} dV \tag{A26}$$

(b) Use phase volume average and derivative of ϕ:

$$\fint_V \phi \frac{\partial \tau_{ij}}{\partial x_j} dV = \frac{\partial}{\partial x_j}\left(\overline{\phi \widetilde{\tau}_{ij}}\right) - \sum_{s=0}^{N_p} \gamma_s \fint_{S_s} \tau_{ij} n_{s,j} dS_s \tag{A27}$$

The final volume-averaged momentum equation reads:

$$\begin{aligned}\frac{\partial}{\partial t}\left(\overline{\phi}\widetilde{\rho}\widehat{u_i}\right) - \fint_V \rho u_i \frac{\partial \phi}{\partial t} dV + \frac{\partial}{\partial x_j}\left(\overline{\phi}\widetilde{\rho}\widehat{u_i u_j} - \overline{\phi}\widetilde{\tau}_{ij}\right) = \\ -\frac{\partial}{\partial x_i}\left(\overline{\phi \widetilde{p}}\right) + \sum_{s=0}^{N_p} \gamma_s \fint_{S_s} \left[p n_{s,i} + \left(\rho u_i u_j - \tau_{ij}\right) n_{s,j}\right] dS_s + \overline{\phi}\widetilde{\rho}\widehat{g}_i\end{aligned} \tag{A28}$$

Introducing the usual "a la Leonard" decomposition:

$$\widetilde{\rho}\widehat{u_i u_j} = \widetilde{\rho}\widehat{u}_i\widehat{u}_j - \tau'_{ij} \tag{A29}$$

And introducing the effective stress tensor:

$$\tau_{ij}^{\text{eff}} = \widetilde{\tau}_{ij} + \tau'_{ij} \tag{A30}$$

We obtain the following:

$$\begin{aligned}\frac{\partial}{\partial t}\left(\overline{\phi}\widetilde{\rho}\widehat{u_i}\right) - \fint_V \rho u_i \frac{\partial \phi}{\partial t} dV + \frac{\partial}{\partial x_j}\left(\overline{\phi}\widetilde{\rho}\,\widehat{u_i}\widehat{u_j} - \overline{\phi}\tau_{ij}^{\text{eff}}\right) = \\ -\frac{\partial}{\partial x_i}\left(\overline{\phi \widetilde{p}}\right) + \sum_{s=0}^{N_p} \gamma_s \fint_{S_s} \left[p n_{s,i} + \left(\rho u_i u_j - \tau_{ij}\right) n_{s,j}\right] dS_s + \overline{\phi}\widetilde{\rho}\widehat{g}_i\end{aligned} \tag{A31}$$

A.2.4 Coarse-grained scalar transport equation

1. Multiply by ϕ and volume average:

$$\fint_V \phi \frac{\partial \theta}{\partial t} dV + \fint_V \phi \frac{\partial}{\partial x_j}(u_j \theta) dV = \fint_V \phi \frac{\partial \Phi_j}{\partial x_j} dV + \fint_V \phi \Theta dV \tag{A32}$$

2. Time derivative

(a) Use integration by parts:

$$\fint_V \phi \frac{\partial \theta}{\partial t} dV = \fint_V \frac{\partial}{\partial t}(\phi \theta) dV - \fint_V \theta \frac{\partial \phi}{\partial t} dV \tag{A33}$$

(b) Use phase volume average:

$$\fint_V \frac{\partial}{\partial t}(\phi\theta)\,dV - \fint_V \theta\frac{\partial\phi}{\partial t}dV = \frac{\partial}{\partial t}\left(\overline{\phi\widetilde{\theta}}\right) - \fint_V \theta\frac{\partial\phi}{\partial t}dV \tag{A34}$$

3. Advection term

(a) Use integration by parts:

$$\fint_V \phi\frac{\partial}{\partial x_j}(u_j\theta)\,dV = \fint_V \frac{\partial}{\partial x_j}(\phi u_j\theta)\,dV - \fint_V (u_j\theta)\frac{\partial\phi}{\partial x_j}dV \tag{A35}$$

(b) Use phase volume average and derivative of ϕ:

$$\fint_V \frac{\partial}{\partial x_j}(\phi u_j\theta)\,dV - \fint_V (u_j\theta)\frac{\partial\phi}{\partial x_j}dV = \frac{\partial}{\partial x_j}\left(\overline{\phi\widetilde{u_j\theta}}\right) - \sum_{s=0}^{N_p}\gamma_s\fint_{S_s}(u_j\theta)\,n_{s,j}dS_s \tag{A36}$$

Applying the same to the diffusion term and the source term one obtains:

$$\begin{aligned}&\frac{\partial}{\partial t}\left(\overline{\phi\widetilde{\theta}}\right) - \fint_V \theta\frac{\partial\phi}{\partial t}dV + \frac{\partial}{\partial x_j}\left(\overline{\phi\widetilde{u_j\theta}} - \overline{\phi\widetilde{\Phi}_j}\right) = \\ &-\sum_{s=0}^{N_p}\gamma_s\fint_{S_s}(u_j\theta + \Phi_j)\,n_{s,j}dS_s + \overline{\phi\widetilde{\Theta}}\end{aligned} \tag{A37}$$

A.2.5 List of coarse-grained transport equations

Mass balance:

$$\frac{\partial}{\partial t}\left(\overline{\phi\widetilde{\rho}}\right) + \frac{\partial}{\partial x_j}\left(\overline{\phi\widetilde{\rho}\widehat{u_j}}\right) = \fint_V \rho\frac{\partial\phi}{\partial t}dV + \sum_{s=0}^{N_p}\gamma_s\fint_{S_s}\rho u_j n_{s,j}dS_s \tag{A38}$$

Momentum conservation:

$$\begin{aligned}&\frac{\partial}{\partial t}\left(\overline{\phi\widetilde{\rho}\,\widehat{u}_i}\right) - \fint_V \rho u_i\frac{\partial\phi}{\partial t}dV + \frac{\partial}{\partial x_j}\left(\overline{\phi\widetilde{\rho}\widehat{u_iu_j}} - \overline{\phi\widetilde{\tau}_{ij}}\right) = \\ &-\frac{\partial}{\partial x_i}\left(\overline{\phi\widetilde{p}}\right) + \sum_{s=0}^{N_p}\gamma_s\fint_{S_s}\left[pn_{s,i} + (\rho u_iu_j - \tau_{ij})\,n_{s,j}\right]dS_s + \overline{\phi\widetilde{\rho}}\,\widehat{g}_i\end{aligned} \tag{A39}$$

Scalar transport:

$$\frac{\partial}{\partial t}\left(\overline{\phi}\widetilde{\theta}\right) - \oint_{V} \theta \frac{\partial \phi}{\partial t} dV + \frac{\partial}{\partial x_j}\left(\overline{\phi}\widetilde{u_j\theta} - \overline{\phi}\widetilde{\Phi}_j\right) = -\sum_{s=0}^{N_p} \gamma_s \oint_{S_s} \left(u_j\theta + \Phi_j\right) n_{s,j} dS_s + \overline{\phi}\widetilde{\Theta} \tag{A40}$$

A.3 Approximations of the volume-averaged governing equations

In the following, we will present some standard approximations to the equations derived (in Section A.2) that allow to significantly simplify their solution process and to disregard some terms that would otherwise require modelling.

A.3.1 No slip condition at particle surfaces

The use of a "no-slip" condition at particle surface results in the fluid velocity field u_i to be equal to the particle velocity v_i at any boundary point. Since v_i is known from the Newton's equations, this condition simplifies all the surface integrals.

A.3.2 Low Mach number (incompressibility limit)

In this regime, the fluid density ρ may be regarded as a constant ρ_0. One important implication is that mass averaged and fluid volume-averaged quantities become identical and there is no distinction between $\widehat{u}_i$ and $\widetilde{u}_i$.

A.3.3 Nondeformable particles

If we consider particles with fixed shape (i.e., not growing or shrinking and unbreakable), then we can simplify the equations by noticing that:

$$\oint_{S_s} \rho_0 v_j n_{s,j} dS_s = 0, \qquad \oint_{S_s} \rho_0 v_i v_j n_{s,j} dS_s = 0 \tag{A41}$$

Which is true if S_s is a closed surface, i.e., if the particle is completely inside V. However, even in case the particle is only partially within V, the previous assumptions ensure that the integrals appear in closed form.

A.3.4 High Stokes number: "gas–particle system"

Integrals containing the time derivative of the indicator function can be shown to vanish in the limit of infinite Stokes number $Stk = \tau_p/\tau_f$, where τ_p is the characteristic time of the dispersed phase and τ_f is the characteristic time of the fluid phase. Since the terms $(\partial\phi)/(\partial t)$ are given by the solution of Newton's equations for the dispersed phase, they are expected to behave like $1/\tau_p$. However, time derivatives of the fluid velocity are behaving like $U_0/\tau_f = U_0^2/d_p$, being U_0 a reference velocity and d_p a characteristic particle diameter. Regardless the specific expression for the Stokes number, in case the particle motion is mainly governed by viscous forces Stk is proportional to the ratio between the particle and fluid density ρ_p/ρ_0. Therefore, terms involving time derivatives of the indicator function are negligible in most of the case where gas–solid suspensions are studied.

A.3.5 Simplified equations

We can therefore write a simplified form of the governing equations which employ the aforementioned assumptions:

Mass balance:

$$\frac{\partial}{\partial t}(\overline{\phi}) + \frac{\partial}{\partial x_j}(\overline{\phi}\widetilde{u_j}) = 0 \tag{A42}$$

Momentum conservation:

$$\begin{aligned} &\frac{\partial}{\partial t}(\overline{\phi}\widetilde{u_i}) + \frac{\partial}{\partial x_j}\left(\overline{\phi}\widetilde{u_i u_j} - \frac{1}{\rho_0}\overline{\phi}\widetilde{\tau_{ij}}\right) = \\ &-\frac{1}{\rho_0}\frac{\partial}{\partial x_i}(\overline{\phi}\widetilde{p}) - \sum_{s=0}^{N_p}\gamma_s \oint_{S_s}\left[\tau_{ij}n_{s,j} - pn_{s,i}\right]dS_s - \overline{\phi}g_i \end{aligned} \tag{A43}$$

Scalar transport:

$$\frac{\partial}{\partial t}(\overline{\phi}\widetilde{\theta}) + \frac{\partial}{\partial x_j}(\overline{\phi}\widetilde{u_j\theta} - \overline{\phi}\widetilde{\Phi_j}) = -\sum_{s=0}^{N_p}\gamma_s \oint_{S_s}(u_j\theta + \Phi_j)\,n_{s,j}dS_s + \overline{\phi}\widetilde{\Theta} \tag{A44}$$

REFERENCES

Abadi M, et al: Tensorflow: Large-scale machine learning on heterogeneous distributed systems: http://download.tensorflow.org/paper/whitepaper2015.pdf, 2015.

Agrawal K, et al: The role of meso-scale structures in rapid gas–solid flows, *J Fluid Mech* 445:151–185, 2001.

Agrawal K, et al: Filtered models for scalar transport in gas–particle flows, *Chem Eng Sci* 95:291–300, 2013.

Akiki G, Jackson TL, Balachandar S: Pairwise interaction extended point-particle model for a random array of monodisperse spheres, *J Fluid Mech* 813:882–928, 2017.

Amini H, Lee W, Di Carlo D: Inertial microfluidic physics, *Lab Chip* 14:2739–2761, 2014.

Anderson TB, Jackson R: A fluid mechanical description of fluidized beds, *Ind Eng Chem Fundam* 6(4):527–539, 1967.
Askarishahi M, Salehi M, Radl S: Full-physics simulations of spray-particle interaction in a bubbling fluidized bed, *AIChE J* 63:2569–2587, 2017.
Baker M, et al: In *RANS modeling of cluster-induced and shear turbulence in particle-laden channel flow*, AIChE annual meet 2016. San Francisco, AIChE, 2016.
Beck D, et al: Data science: accelerating innovation and discovery in chemical engineering, *AIChE J* 62(5):1402–1416, 2016.
Beetstra R, van der Hoef MA, Kuipers JAM: Drag force of intermediate Reynolds number flow past mono- and bidisperse arrays of spheres, *AIChE J* 53(2):489–501, 2007.
Berthold M, et al: KNIME: the Konstanz information miner, *ACM SIGKDD Explor Newsl* 11(1):26–31, 2008.
Bierwisch C, et al: Three-dimensional discrete element models for the granular statics and dynamics of powders in cavity filling, *J Mech Phys Solids* 57(1):10–31, 2009.
Boyce CM, et al: Growth and breakup of a wet agglomerate in a dry gas-solid fluidized bed, *AIChE J* 63:2520–2527, 2017.
Brady JF, Bossis G: Stokesian dynamics, *Annu Rev Fluid Mech* 20(1):111–157, 1988.
Buist KA, et al: A combined experimental and simulation study of fluid-particle heat transfer in dense arrays of stationary particles, *Chem Eng Sci* 169:310–320, 2017.
Calzavarini E, et al: Quantifying turbulence induced segregation of inertial particles, *Phys Rev Lett* 101(2):84504, 2008.
Capecelatro J, Desjardins O: An Euler–Lagrange strategy for simulating particle-laden flows, *J Comput Phys* 238:1–31, 2013.
Capecelatro J, Desjardins O, Fox RO: Numerical study of collisional particle dynamics in cluster-induced turbulence, *J Fluid Mech* 747:R2, 2014.
Capecelatro J, Desjardins O, Fox RO: On fluid–particle dynamics in fully developed cluster-induced turbulence, *J Fluid Mech* 780:578–635, 2015.
Capecelatro J, Desjardins O, Fox RO: Strongly coupled fluid-particle flows in vertical channels. II. Turbulence modeling, *Phys Fluids* 28(3):33307, 2016.
Cloete S, Amini S, Johansen ST: On the effect of cluster resolution in riser flows on momentum and reaction kinetic interaction, *Powder Technol* 210:6–17, 2011.
Cloete S, Johansen ST, Amini S: Performance evaluation of a complete Lagrangian KTGF approach for dilute granular flow modelling, *Powder Technol* 226:43–52, 2012.
Cloete JH, et al: Evaluation of wall friction models for riser flow, *Powder Technol* 303:156–167, 2016.
Cloete J.H., et al: An anisotropic mesoscale solids stress closure for freely sedimenting gas-particle suspensions, *Powder Technol* 2017a, under revi.
Cloete JH, et al: The sensitivity of filtered two fluid model to the underlying resolved simulation setup, *Powder Technol* 316:265–277, 2017a.
Cloete JH, et al: In Olsen JEO, Johansen ST, editors: *Verification of filtered two fluid models for reactive gas-solid flows*, 12th international conference on CFD oil gas, metall process industries. Trondheim, Norway, SINTEF, Norway, 2017b.
Cundall PA, Strack ODL: A discrete numerical model for granular assemblies, *Géotechnique* 47–65, 1979.
de Baas AF, Rosso L: *Modelling in FP7 NMP programme materials projects,* Brussels, Belgium, 2015, Brussels, Belgium, European Commission, Directorate-General for Research and Innovation.
de Klerk A: Voidage variation in packed beds at small column to particle diameter ratio, *AIChE J* 49(8):2022–2029, 2003.
Deen NG, Kuipers JAM: Direct numerical simulation (DNS) of mass, momentum and heat transfer in dense fluid-particle systems, *Curr Opin Chem Eng* 5:84–89, 2014a.
Deen NG, Kuipers JAM: Direct numerical simulation of fluid flow accompanied by coupled mass and heat transfer in dense fluid–particle systems, *Chem Eng Sci* 116:645–656, 2014b.

Deen NG, et al: Direct numerical simulation of flow and heat transfer in dense fluid-particle systems, *Chem Eng Sci* 81:329–344, 2012.
Deen NG, et al: Review of direct numerical simulation of fluid-particle mass, momentum and heat transfer in dense gas-solid flows, *Chem Eng Sci* 116:710–724, 2014.
Delgado JMPQ: A critical review of dispersion in packed beds, *Heat Mass Transf* 42:279–310, 2006.
Derksen JJ: Mixing by solid particles, *Chem Eng Res Des* 86(12):1363–1368, 2008a.
Derksen JJ: Scalar mixing by granular particles, *AIChE J* 54(7):1741–1747, 2008b.
Derksen JJ: Scalar mixing with fixed and fluidized particles in micro-reactors, *Chem Eng Res Des* 87(4):550–556, 2009.
Derksen JJ: Simulations of scalar dispersion in fluidized solid–liquid suspensions, *AIChE J* 60(5):1880–1890, 2014a.
Derksen JJ: Simulations of solid–liquid mass transfer in fixed and fluidized beds, *Chem Eng J* 255:233–244, 2014b.
Derksen JJ, Sundaresan S: Direct numerical simulations of dense suspensions: wave instabilities in liquid-fluidized beds, *J Fluid Mech* 587:303–336, 2007.
Desjardins O, Fox RO, Villedieu P: A quadrature-based moment method for dilute fluid-particle flows, *J Comput Phys* 227(4):2514–2539, 2008.
Di Carlo D: Inertial microfluidics, *Lab Chip* 9(21):3038–3046, 2009.
Di Carlo D, et al: Particle segregation and dynamics in confined flows, *Phys Rev Lett* 102(9):94503, 2009.
Ding J, Gidaspow D: A bubbling fluidization model using kinetic-theory of granular flow, *AIChE J* 36(4):523–538, 1990.
Dixon AG: Local transport and reaction rates in a fixed bed reactor tube: endothermic steam methane reforming, *Chem Eng Sci* 168:156–177, 2017.
Domingo P, Vervisch L: DNS and approximate deconvolution as a tool to analyse one-dimensional filtered flame sub-grid scale modelling, *Combust Flame* 177:109–122, 2017.
Egholm DL: A new strategy for discrete element numerical models: 1. Theory, *J Geophys Res Earth* 112(B5), 2007.
Egholm DL, et al: A new strategy for discrete element numerical models: 2. Sandbox applications, *J Geophys Res Earth* 112(B5), 2007.
Eswaran V, Pope SB: An examination of forcing in direct numerical simulations of turbulence, *Comput Fluids* 257–278, 1988.
Fan FG, Ahmadi G: A sublayer model for wall deposition of ellipsoidal particles in turbulent streams, *J Aerosol Sci* 26(5):813–840, 1995.
Feng ZG, Michaelides EE: Heat transfer in particulate flows with direct numerical simulation (DNS), *Int J Heat Mass Transf* 52(3–4):777–786, 2009.
Feng Z-G, Musong SG: Direct numerical simulation of heat and mass transfer of spheres in a fluidized bed, *Powder Technol* 262:62–70, 2014.
Forgber T, Radl S: A novel approach to calculate radiative thermal exchange in coupled particle simulations, *Powder Technol* 323:24–44, 2018.
Forgber T, et al: Heat transfer rates in sheared beds of inertial particles at high Biot numbers, *Granul Matter* 19(14), 2017.
Fotovat F, Bi XT, Grace JR: Electrostatics in gas-solid fluidized beds: a review, *Chem Eng Sci* 173:303–334, 2017.
Fox RO: A quadrature-based third-order moment method for dilute gas-particle flows, *J Comput Phys* 227(12):6313–6350, 2008.
Fox RO: Large-Eddy-simulation tools for multiphase flows, *Annu Rev Fluid Mech* 44:47–76, 2012.
Fox RO: In *Quadrature-based moment methods for fluid-particle flows*, Part Technol Forum (Shell Thomas Baron Award Fluid-Particle System), 2017, pp 28–40.
Fox RO, et al: In *Clustering in gas-solid flows: how are clusters modified by shear?* AIChE annual meet 2016. San Francisco, AIChE, 2016.

Fullmer WD, Hrenya CM: Quantitative assessment of fine-grid kinetic-theory-based predictions of mean-slip in unbounded fluidization, *AIChE J* 62:11–17, 2016.
Fullmer WD, Hrenya CM: The clustering instability in rapid granular and gas-solid flows, *Annu Rev Fluid Mech* 49:485–510, 2017.
Galvin JE, Benyahia S: The effect of cohesive forces on the fluidization of aeratable powders, *AIChE J* 60(2):473–484, 2014.
Ganguli S, Lele S: In *Drag of spherical particles in a periodic lattice: heat transfer, buoyancy and non-boussinesq effect*, APS DFD meet 2016. Portland, Oregon, American Physical Society, 2016.
Gidaspow D: *Multiphase flow and fluidization: continuum and kinetic theory descriptions,* San Diego, Califonria, U.S.A, 1994, Academic Press.
Girardi M, Radl S, Sundaresan S: In *Liquid distribution in wet fluidized beds*, AIChE annual meet 2012. Pittsburgh, PA, United States, AIChE, 2012.
Girardi M, Radl S, Sundaresan S: Simulating wet gas-solid fluidized beds using coarse-grid CFD-DEM, *Chem Eng Sci* 144:224–238, 2016.
Glasser BJ, Sundaresan S, Kevrekidis I: From bubbles to clusters in fluidized beds, *Phys Rev Lett* 81(9):1849–1852, 1998.
Goniva C, et al: Influence of rolling friction on single spout fluidized bed simulation, *Particuology* 10(5):582–591, 2012.
Greidinger Z, et al: In *Agglomerate size distribution in wet gas fluidized systems*, AIChE annual meet 2014. Atlanta, U.S., AIChE, 2014.
Gu Y, Ozel A, Sundaresan S: A modified cohesion model for CFD-DEM simulations of fluidization, *Powder Technol* 296:17–28, 2016a.
Gu Y, Ozel A, Sundaresan S: Numerical studies of the effects of fines on fluidization, *AIChE J* 62(7):2271–2281, 2016b.
Gunn DJ: Transfer of heat or mass to particles in fixed and fluidised beds, *Int J Heat Mass Transf* 21:467–476, 1978.
Hærvig J, et al: On the adhesive JKR contact and rolling models for reduced particle stiffness discrete element simulations, *Powder Technol* 319:472–482, 2017.
Hagelien T, et al: In Kilappa V, Matilla K, editors: *Porto: a framework for information interchange and multi-scale fluid mechanics simulations*, 3rd work European multi-scale model cluster. Jyväskylä, Finland, University of Jyväskylä, Finland, 2015, pp 1–33.
Holloway WS: *Continuum modeling of gas-particle flows across multiple scales,* PhD Thesis, 2012, Princeton University.
Holloway WS, Sundaresan S: In *Filtered models for reacting gas-solid flows*, AIChE annual meet 2011. AIChE, 2011.
Holloway WS, Sundaresan S: Filtered models for reacting gas–particle flows, *Chem Eng Sci* 82:132–143, 2012.
Holloway WS, Sundaresan S: Filtered models for bidisperse gas–particle flows, *Chem Eng Sci* 108:67–86, 2014.
Holloway WS, Yin X, Sundaresan S: Fluid–particle drag in inertial polydisperse gas–solid suspensions, *AIChE J* 56(8):1995–2004, 2010.
Holloway WS, et al: Meso-scale structures of bidisperse mixtures of particles fluidized by a gas, *Chem Eng Sci* 66(19):4403–4420, 2011.
Holloway WS, Sun J, Sundaresan S: Effect of microstructural anisotropy on the fluid-particle drag force and the stability of the uniformly fluidized state, *J Fluid Mech* 713:27–49, 2012.
Hölzer A, Sommerfeld M: Lattice Boltzmann simulations to determine drag, lift and torque acting on non-spherical particles, *Comput Fluids* 38(3):572–589, 2009.
Horwitz J, et al: In *A direct comparison of fully resolved and point-particle models in particle-laden turbulent flow*, APS DFD meet 2016. Portland, Oregon, American Physical Society, 2016.
Iddir H, Arastoopour H: Modeling of multitype particle flow using the kinetic theory approach, *AIChE J* 51(6):1620–1632, 2005.
Igci Y, Sundaresan S: Constitutive models for filtered two-fluid models of fluidized gas-particle flows, *Ind Eng Chem Res* 50:13190–13201, 2011.

Igci Y, et al: Filtered two-fluid models for fluidized gas-particle suspensions, *AIChE J* 54(6):1431–1448, 2008.

Jackson R: *The dynamics of fluidized particles,* Cambridge, UK, 2000, Cambridge University Press (Cambridge Monographs on Mechanics).

Jajcevic D, et al: Large-scale CFD-DEM simulations of fluidized granular systems, *Chem Eng Sci* 98:298–310, 2013.

Johnson PC, Jackson R: Frictional–collisional constitutive relations for granular materials, with application to plane shearing, *J Fluid Mech* 176:67–93, 1987.

Kloss C, et al: Models, algorithms and validation for opensource DEM and CFD-DEM, *Prog Comput Fluid Dyn* 12:140–152, 2012.

Kolakaluri R: *Direct numerical simulations and analytical modeling of granular filtration,* 2013, PhD Thesis, Iowa State University.

Kolakaluri R, et al: Filtration model for polydisperse aerosols in gas-solid flow using granule-resolved direct numerical simulation, *AIChE J* 61(11):3594–3606, 2015.

Kolehmainen J, et al: A hybrid approach to computing electrostatic forces in fluidized beds of charged particles, *AIChE J* 62(7):2282–2295, 2016.

Kolehmainen J, Sippola P, et al: Effect of humidity on triboelectric charging in a vertically vibrated granular bed: experiments and modeling, *Chem Eng Sci* 173:363–373, 2017a.

Kolehmainen J, Ozel A, et al: Triboelectric charging of monodisperse particles in fluidized beds, *AIChE J* 63(6):1872–1891, 2017b.

Kolehmainen J, Ozel A, Sundaresan S: In *Gas-solid flows with tribocharging*, AIChE annual meet 2017. Minneapolis, USA, AIChE, 2017c.

Kong B, Fox RO: A solution algorithm for fluid–particle flows across all flow regimes, *J Comput Phys* 344:575–594, 2017.

Kong B, et al: Euler–Euler anisotropic Gaussian mesoscale simulation of homogeneous cluster-induced gas–particle turbulence, *AIChE J* 63:2630–2643, 2017.

Kravets B, Kruggel-Emden H: Investigation of local heat transfer in random particle packings by a fully resolved LBM-approach, *Powder Technol* 318:293–305, 2017.

Kriebitzsch SHL, van der Hoef MA, Kuipers JAM: Drag force in discrete particle models—continuum scale or single particle scale? *AIChE J* 59(1):316–324, 2013.

Kuerten JGM: Subgrid modeling in particle-laden channel flow, *Phys Fluids* 18(2):25108, 2006.

Ladd AJC: Numerical simulations of particulate suspensions via a discretized Boltzmann equation. Part 1. Theoretical foundation, *J Fluid Mech* 271:285–309, 1994.

Lane WA, et al: Sub-grid models for heat transfer in gas-particle flows with immersed horizontal cylinders, *Chem Eng Sci* 151:7–15, 2016.

Lees AW, Edwards SF: The computer study of transport processes under extreme conditions, *J Phys Chem* 5:1921–1930, 1972.

Lekhal A, Glasser BJ, Khinast JG: Impact of drying on the catalyst profile in supported impregnation catalysts, *Chem Eng Sci* 56(15):4473–4487, 2001a.

Lekhal A, Khinast JG, Glasser BJ: Predicting the effect of drying on supported coimpregnation catalysts, *Ind Eng Chem Res* 40(18):3989–3999, 2001b.

Lu G, Third JR, Müller CR: Discrete element models for non-spherical particle systems: from theoretical developments to applications, *Chem Eng Sci* 127:425–465, 2015.

Ma T, et al: Kernel-assisted and topology-aware MPI collective communications on multicore/many-core platforms, *J Parallel Distrib Comput* 73(7):1000–1010, 2013.

Maheshwari A, Chhabra RP, Biswas G: Effect of blockage on drag and heat transfer from a single sphere and an in-line array of three spheres, *Powder Technol* 168(2):74–83, 2006.

Mahmoudi AH, et al: Modeling of the biomass combustion on a forward acting grate using XDEM, *Chem Eng Sci* 142:32–41, 2016.

Mamidala AR, et al: In *MPI collectives on modern multicore clusters: performance optimizations and communication characteristics*, Proceedings of CCGRID 2008—8th IEEE international symposium on cluster computing grid, 2008, pp 130–137.

Marchioli C: Large-eddy simulation of turbulent dispersed flows: a review of modelling approaches, *Acta Mech* 228(3):741–771, 2017.

Marchisio DL, Fox RO: *Computational models for polydisperse particulate and multiphase systems,* Cambridge, UK, 2013, Cambridge University Press.

Milioli CC, et al: Filtered two-fluid models of fluidized gas-particle flows: new constitutive relations, *AIChE J* 59:3265–3275, 2013.

Moureau V, Domingo P, Vervisch L: From large-eddy simulation to direct numerical simulation of a lean premixed swirl flame: filtered laminar flame-PDF modeling, *Combust Flame* 158(7):1340–1357, 2011.

Mueller GE: Radial void fraction distributions in randomly packed fixed beds of uniformly sized spheres in cylindrical containers, *Powder Technol* 72:269–275, 1992.

Municchi F, Radl S: Consistent closures for Euler-Lagrange models of bi-disperse gas-particle suspensions derived from particle-resolved direct numerical simulations, *Int J Heat Mass Transf* 111:171–190, 2017.

Municchi F, Radl S: Momentum, heat and mass transfer simulations of bounded dense mono-dispersed gas-particle systems, *Int J Heat Mass Transf* 120:1146–1161, 2018.

Municchi F, Goniva C, Radl S: Highly efficient spatial data filtering in parallel using the opensource library CPPPO, *Comput Phys Commun* 207:400–414, 2016.

NanoSim—a multiscale simulation-based design platform for cost-effective CO2 capture processes using nano-structured materials: https://www.sintef.no/projectweb/nanosim/, 2017.

Nakariyakul S: Fast spatial averaging: an efficient algorithm for 2D mean filtering, *J Supercomput* 65(1):262–273, 2013.

Oschmann T, Schiemann M, Kruggel-Emden H: Development and verification of a resolved 3D inner particle heat transfer model for the discrete element method (DEM), *Powder Technol* 291:392–407, 2016.

Ouchene R, et al: In *Numerical simulation and modelling of the forces acting on single and multiple non-spherical particles*, Proceeding ASME 2014 4th Jt US-European Fluids Eng. Chicago, Illinois, USA, ASME, 2014.

Ouchene R, et al: Drag, lift and torque coefficients for ellipsoidal particles: from low to moderate particle Reynolds numbers, *Comput Fluids* 113:53–64, 2015.

Ouchene R, et al: A new set of correlations of drag, lift and torque coefficients for non-spherical particles and large Reynolds numbers, *Powder Technol* 303:33–43, 2016.

Ozel A, Fede P, Simonin O: Development of filtered Euler–Euler two-phase model for circulating fluidised bed: high resolution simulation, formulation and a priori analyses, *Int J Multiphase Flow* 55:43–63, 2013.

Ozel A, et al: Fluid and particle coarsening of drag force for discrete-parcel approach, *Chem Eng Sci* 155:258–267, 2016.

Ozel A, et al: Towards fitered drag force model for non-cohesive and cohesive particle-gas flows, *Phys Fluids* 29, 2017, 103308.

Padding JT: In Padding JT, editor: *Lecture notes for the course on particle-based simulations*, Delft, 2013, Delft University of Technology.

Pannala S, Syamlal M, O'Brien TJ, O'Brien TJ, Hershey PA: In Pannala S, Syamlal M, editors: *Computational gas-solids flows and reacting systems: theory, methods and practice*, United States, 2011, Engineering Science Reference.

Parmentier J-F, Simonin O, Delsart O: A functional subgrid drift velocity model for filtered drag prediction in dense fluidized bed, *AIChE J* 58:1084–1098, 2011.

Passalacqua A, Fox RO: Advanced continuum modelling of gas-particle flows beyond the hydrodynamic limit, *App Math Model* 35(4):1616–1627, 2011.

Patankar NA, Joseph DD: Modeling and numerical simulation of particulate flows by the Eulerian-Lagrangian approach, *Int J Multiphase Flow* 27(10):1659–1684, 2001.

Pirker S, Kahrimanovic D, Aichinger G: Modeling mass loading effects in industrial cyclones by a combined Eulerian-Lagrangian approach, *Acta Mech* 204(3–4):203–216, 2009.

Pirker S, et al: Application of a hybrid lattice Boltzmann-finite volume turbulence model to cyclone short-cut flow, *Powder Technol* 235:572–580, 2013.
Podlozhnyuk A, et al: In *Modelling superquadric particles using hybrid fictitious domain-immersed boundary method*, 11th OpenFOAM® Work. Guimaraes, Portugal, University of Minho, Institute of Polymers and Composites, Portugal, 2016.
Popoff B, Braun M: In *A Lagrangian approach to dense particulate flows*, 6th international conference on multiphase flow, 2007, p 510.
Radeke C, Glasser BJ, Khinast JG: Large-scale powder mixer simulations using massively parallel GPU architectures, *Chem Eng Sci* 65(24):6435–6442, 2010.
Radl S, Sundaresan S: A drag model for filtered Euler-Lagrange simulations of clustered gas-particle suspensions, *Chem Eng Sci* 117:416–425, 2014.
Radl S, et al: In *Parcel-based approach for the simulation of gas-particle flows*, 8th International conference CFD oil gas, metal process industries. Trondheim, Norway, SINTEF/NTNU Trondheim, Norway124, 2011, pp 1–10.
Radl S, Girardi M, Sundaresan S: In Eberhardsteiner J, editor: *Effective drag law for parcel-based approaches—what can we learn from CFD-DEM?* ECCOMAS 2012—European congress computational methods in applied sciences engineering. Vienna, Austria, 2012, pp 1225–1239.
Radl S, Krainer F, et al: In *Biot number effects on the local heat and mass transfer rate in fixed and fluidized beds*, AIChE annual meet 2015. Salt Lake City, U.S.A., AIChE, 2015a.
Radl S, Forgber T, et al: In Onate E, et al, editor: *ParScale—an open-source library for the simulation of intra-particle heat and mass transport processes in coupled simulations*, IV International conference part methods—fundamental application (PARTICLES 2015). Barcelona, Spain, ECCOMAS1–9, 2015b.
Rantanen J, Khinast J: The future of pharmaceutical manufacturing sciences, *J Pharm Sci* 104(11):3612–3638, 2015.
Rhodes M, et al: Onset of cohesive behaviour in gas fluidized beds: a numerical study using DEM simulation, *Chem Eng Sci* 56(14):4433–4438, 2001.
Roadmap chemical reaction engineering: http://dechema.de/dechema_media/Reaktionstechnik_Roadmap_2017_en-p-20002685.pdf, 2017.
Rohde M, Derksen JJ, Van Den Akker HEA: Volumetric method for calculating the flow around moving objects in lattice-Boltzmann schemes, *Phys Rev E* 65(5):56701, 2002.
Rourke PJO, Snider DM: A new blended acceleration model for the particle contact forces induced by an interstitial fluid in dense particle/fluid flows, *Powder Technol* 256:39–51, 2014.
Rubinstein GJ, Derksen JJ, Sundaresan S: Lattice Boltzmann simulations of low-Reynolds-number flow past fluidized spheres: effect of Stokes number on drag force, *J Fluid Mech* 788:576–601, 2016.
Sagaut P: *Large Eddy simulation for incompressible flows—an introduction*, ed 3, Berlin, Germany, 2006, Springer, Berlin.
Sakai M, Koshizuka S: Large-scale discrete element modeling in pneumatic conveying, *Chem Eng Sci* 64(3):533–539, 2009.
Sarkar A, et al: Filtered sub-grid constitutive models for fluidized gas-particle flows constructed from 3-D simulations, *Chem Eng Sci* 152:443–456, 2016.
Schellander D, Schneiderbauer S, Pirker S: Numerical study of dilute and dense poly-dispersed gas-solid two-phase flows using an Eulerian and Lagrangian hybrid model, *Chem Eng Sci* 95:107–118, 2013.
Schmidt R, Nikrityuk PA: Numerical simulation of the transient temperature distribution inside moving particles, *Can J Chem Eng* 90(2):246–262, 2012.
Schneiderbauer S: A spatially-averaged two-fluid model for dense large-scale gas-solid flows, *AIChE J* 63(8):3544–3562, 2017.
Schneiderbauer S, Pirker S: Filtered and heterogeneity-based subgrid modifications for gas–solid drag and solid stresses in bubbling fluidized beds, *AIChE J* 60(3):839–854, 2014.

Schneiderbauer S, Saeedipour M: Approximate deconvolution model for the simulation of turbulent gas-solid flows: an a-priori analysis, *Phys Fluids* 30:23301, 2018.
Schneiderbauer S, et al: Non-steady state boundary conditions for collisional granular flows at flat frictional moving walls, *Int J Multiphase Flow* 43:149–156, 2012a.
Schneiderbauer S, Aigner A, Pirker S: A comprehensive frictional-kinetic model for gas–particle flows: analysis of fluidized and moving bed regimes, *Chem Eng Sci* 80:279–292, 2012b.
Schneiderbauer S, et al: A Lagrangian-Eulerian hybrid model for the simulation of industrial-scale gas-solid cyclones, *Powder Technol* 304:229–240, 2016.
Shinbrot T, Rutala M, Herrmann H: Surface contact charging, *Phys Rev E* 96(3):32912, 2017.
Sierou A, Brady JF: Accelerated Stokesian dynamics simulations, *J Fluid Mech* 448:115–146, 2001.
Singhal A, et al: Heat transfer to a gas from densely packed beds of monodisperse spherical particles, *Chem Eng J* 314:27–37, 2017.
Snider DM: An incompressible three-dimensional multiphase particle-in-cell model for dense particle flows, *J Comput Phys* 170(2):523–549, 2001.
Sommerfeld M: Analysis of collision effects for turbulent gas-particle flow in a horizontal channel: part 1. Particle transport, *Int J Multiphase Flow* 29(4):675–699, 2003.
Stan CA, et al: The magnitude of lift forces acting on drops and bubbles in liquids flowing inside microchannels, *Lab Chip* 13(3):365–376, 2013.
Stark AK, et al: Towards an advanced reactor network modeling framework for fluidized bed biomass gasification: incorporating information from detailed CFD simulations, *Chem Eng J* 303:409–424, 2016.
Sun B, Tenneti S, Subramaniam S: Modeling average gas-solid heat transfer using particle-resolved direct numerical simulation, *Int J Heat Mass Transf* 86:898–913, 2015.
Sundaresan S: Modeling the hydrodynamics of multiphase flow reactors: current status and challenges, *AIChE J* 46(6):1102–1105, 2000.
Sundaresan S: Instabilities in fluidized beds, *Annu Rev Fluid Mech* 35(1):63–88, 2003.
Sundaresan S, et al: Coarse-grained models for momentum, energy and species transport in gas-particle flows. In *Proceedings of the 14th international conference on fluidization—from fundamentals to products, 2013, Noordwijkerhout, The Netherlands*, 2013, Engineering Conferences International.
Tang Y, Peters EAJF, Kuipers JAM: Direct numerical simulations of dynamic gas-solid suspensions, *AIChE J* 62(6):1958–1969, 2016.
Tavassoli H, et al: Direct numerical simulation of particulate flow with heat transfer, *Int J Multiphase Flow* 57:29–37, 2013.
Tavassoli H, Peters EAJF, Kuipers JAM: Direct numerical simulation of fluid-particle heat transfer in fixed random arrays of non-spherical particles, *Chem Eng Sci* 129:42–48, 2015.
Ten Cate A, et al: Fully resolved simulations of colliding monodisperse spheres in forced isotropic turbulence, *J Fluid Mech* 519:233–271, 2004.
Tenneti S, Subramaniam S: Particle-resolved direct numerical simulation for gas-solid flow model development, *Annu Rev Fluid Mech* 46(1):199–230, 2014.
Tenneti S, et al: Role of fluid heating in dense gas-solid flow as revealed by particle-resolved direct numerical simulation, *Int J Heat Mass Transf* 58:471–479, 2013.
Theuerkauf J, Witt P, Schwesig D: Analysis of particle porosity distribution in fixed beds using the discrete element method, *Powder Technol* 165:92–99, 2006.
Tsai CM, Yeh ZM: Fast spatial averaging filter, international symposium on computer, consumer and control, 1:153–156, 2012.
Uhlmann M, Doychev T: Sedimentation of a dilute suspension of rigid spheres at intermediate Galileo numbers: the effect of clustering upon the particle motion, *J Fluid Mech* 752:310–348, 2014.

van Antwerpen W, Toit CG, Rousseau PG: A review of correlations to model the packing structure and effective thermal conductivity in packed beds of mono-sized spherical particles, *Nucl Eng Des* 240(7):1803–1818, 2010.
van der Hoef MA, Beetstra R, Kuipers JAM: Lattice-Boltzmann simulations of low-Reynolds-number flow past mono- and bidisperse arrays of spheres: results for the permeability and drag force, *J Fluid Mech* 528:233–254, 2005.
van der Hoef MA, et al: Numerical simulation of dense gas-solid fluidized beds: a multiscale modeling strategy, *Annu Rev Fluid Mech* 40(1):47–70, 2008.
Wen C, Yu Y: Mechanics of fluidization, *Chem Eng Prog Symp Ser* 62:100–111, 1966.
Whitaker S: *The method of volume averaging,* Dordrecht, NL, 1999, Springer Science + Business Media.
Wu M, Khinast JG, Radl S: In *Direct simulation of film flow on spheres to investigate liquid bridge,* AIChE annual meet 2014. Atlanta, Vereinigte Staaten (USA), AIChE, 2014.
Wu M, Khinast JG, Radl S: A model to predict liquid bridge formation between wet particles based on direct numerical simulations, *AIChE J* 62(6):1877–1897, 2016.
Wu M, Khinast JG, Radl S: Liquid transport rates during binary collisions of unequally-sized particles, *Powder Technol* 309:95–109, 2017.
Wu M, Khinast JG, Radl S: The effect of liquid bridge model details on the dynamics of wet fluidized beds, *AIChE J* 64:437–456, 2018.
Wurzenberger JC, et al: Thermal conversion of biomass: comprehensive reactor and particle modeling, *AIChE J* 48(10):2398–2411, 2002.
Xiong Q, et al: SPH method for two-fluid modeling of particle–fluid fluidization, *Chem Eng Sci* 66:1859–1865, 2011.
Ye M, van der Hoef MA, Kuipers JAM: A numerical study of fluidization behavior of Geldart A particles using a discrete particle model, *Powder Technol* 139(2):129–139, 2004.
Yin X, Sundaresan S: Fluid-particle drag in low-Reynolds-number Polydisperse gas-solid suspensions, *AIChE J* 55(6):1352–1368, 2009.
Yuan C, Fox RO: Conditional quadrature method of moments for kinetic equations, *J Comput Phys* 230(22):8216–8246, 2011.
Yuu S, Umekage T: Constitutive relations and computer simulation of granular material, *Adv Powder Technol* 19(3):203–230, 2008.
Zaidi AA, Tsuji T, Tanaka T: A new relation of drag force for high Stokes number monodisperse spheres by direct numerical simulation, *Adv Powder Technol* 25(6):1860–1871, 2014.
Zastawny M, et al: Derivation of drag and lift force and torque coefficients for non-spherical particles in flows, *Int J Multiph Flow* 39:227–239, 2012.
Zhou Q, Wang J: CFD study of mixing and segregation in CFB risers: extension of EMMS drag model to binary gas-solid flow, *Chem Eng Sci* 122:637–651, 2015.
Zhou ZY, et al: Discrete particle simulation of particle–fluid flow: model formulations and their applicability, *J Fluid Mech* 661:482–510, 2010.
Zhu L-T, Ye M, Luo Z: Application of filtered model for reacting gas – solid flows and optimization in a large-scale methanol-to-olefin fluidized-bed reactor, *Ind Eng Chem Res* 55:11887–11899, 2016.
Zick A, Homsy G: Stokes flow through periodic arrays of spheres, *J Fluid Mech* 115:13–26, 1982.

INDEX

Note: Page numbers followed by "*f*" indicate figures, and "*t*" indicate tables.

E

F

Kolibri 360 GmbH
Ferdinand-Jühlke-Straße 7
99095 Erfurt, Deutschland
produktsicherheit@kolibri360.de